Nonthermal Mechanism of Low-Level Microwave Effect on the Human Brain

This book presents a systematic description of the physical mechanisms and properties of low-level microwave radiation effects. It introduces the concept of nonthermal mechanism of microwave radiation effects, with special reference to the nonlinearity of neurophysiological and biological processes. With insights into electromagnetism, bioelectromagnetism, and nonlinear dynamics, this comprehensive volume proposes new ideas for updating the health protection limits for radiofrequency radiation and recommends following the precautionary principle to minimize health risks.

Features:

- Explains how cumulative minor disturbances from coherent microwave radiation overcome thermal noise limits, enabling low-level microwave impact on biological processes.
- Confirms that dielectric polarization triggers small initial changes that produce significant alterations in nonlinear living systems, causing nonthermal effects.
- Demonstrates that predominantly positive results across multiple tissues and cells from numerous independent studies support the capability of low-level radiation to cause biological effect.
- Shows why modulation frequency is crucial in microwave effects, with the impact directly dependent on modulation frequency.
- Concludes how brain effects vary between individuals, with only a minority experiencing significant impact.
- Emphasizes that the body's repair mechanisms prevent biological effects from becoming health effects, due to nonlinear chaotic nature of the processes with unpredictable outcomes.

This book is aimed at graduate students, researchers, specialists, and academics in biomedical engineering, physics, medicine, medical physics, microwave engineering, health protection, biophysics, and biochemistry.

Nonthermal Mechanism of Low-Level Microwave Effect on the Human Brain

Hiie Hinrikus

CRC Press
Taylor & Francis Group
Boca Raton London New York

CRC Press is an imprint of the
Taylor & Francis Group, an **informa** business

Cover: Shutterstock

First edition published 2026
by CRC Press
2385 NW Executive Center Drive, Suite 320, Boca Raton FL 33431

and by CRC Press
4 Park Square, Milton Park, Abingdon, Oxon, OX14 4RN

CRC Press is an imprint of Taylor & Francis Group, LLC

ISBN: 9781041219811 (hbk)
ISBN: 9781041219828 (pbk)
ISBN: 9781003729976 (ebk)

DOI: 10.1201/9781003729976

Typeset in Times
by codeMantra

Contents

Preface

This book delivers an important contribution to our understanding of how weak electromagnetic fields in the microwave spectrum influence biological systems.

There is a confusing situation regarding the effect and health impact of low-level microwave radiation today. People are living in low-level microwave fields worldwide and are worried about the health effects of the radiation. During the last decades, the majority of studies have reported the biological effects of low-level microwave radiation. However, the legislation on health protection limits is based on thermal effects caused by high-level microwave radiation and relatively short-term exposures. The impact of low-level radiation and long-term exposures, ranging from hours to years, is not taken into account. A possible reason is that there is still no mechanism of low-level microwave radiation commonly accepted by all scientists. The low-level microwave effects have been reported by multiple investigators, but their possible health effects are still debated. This book contributes to solving this problem.

This book is based on the principles of electromagnetism, bioelectromagnetism, and nonlinear dynamics, which are applied to explain the physical mechanism of low-level microwave radiation. This book demonstrates that dielectric polarization of tissues by coherent electromagnetic radiation can lead to biological effects.

The author explains the effects on polarization of molecules with dipole moments upon the application of electromagnetic fields, showing that the viscosity of these molecules moving through a biological fluid is reduced, and the mobility increases. This, in turn, reduces the time required to change the concentrations of important biological signaling molecules to reach threshold levels for initiating changes in function at locations such as synapses. These increases in mobility are not predicted to have an electromagnetic threshold for biological effects. The importance of the changes in viscosity is not widely recognized by many workers in the field.

This book highlights the nonlinear effects introduced by the active nature of living systems that enable small perturbations to initiate significant biological changes, particularly in the brain. It clearly shows that signals from cell phones are sensed by the brain and modify the EEG. The related biological mechanism remains unknown regarding how these signals affect brain neurobiology and how the controlling brain signals might be used to treat diseases.

This book is a good starting point for individuals entering the field, particularly those with a background in biology or medicine, and who are interested in gaining an understanding of the physical mechanisms underlying the ability of biological systems to detect weak microwave electromagnetic fields and some properties of their effects on these systems. For those concerned about possible health effects due to the large variety of sources we are all exposed to these days, from cell phones, WiFi, microwave ovens, smart meters, etc., a good introduction to the large number of papers and reviews is provided. It is balanced in showing both data that show biological effects and those that show inconclusive or conflicting results. The variation in these results is predicted to be, in large part, due to the variability and complexity of biological systems and the exposure parameters. The selection criteria for acceptable articles by different reviewers also affect their conclusions on the importance of weak microwave exposure on possible health effects.

Over the coming years, we will likely get a more complete understanding of the effects of microwave exposures and the environmental conditions that modify the biological effects of these weak fields. We are also likely to gain a better understanding of safety conditions and how they vary in relation to environmental factors, the age, and the health status of exposed individuals. We can also see that our understanding of these effects will help specify exposure parameters for the treatment of a wide variety of diseases.

Frank Barnes
Colorado, USA
July 2025

Acknowledgments

My deep gratitude to my long-term colleagues Jaanus Lass and Maie Bachmann, without whose enthusiasm, dedication, and experience in signal processing, this book would not have been possible. I am very grateful to Jüri Engelbrecht for introducing me to the nonlinear world. Many thanks to my colleagues Anna Suhhova-Jämsa, Jevgeni Riipulk, Anastasia Rodina-Theocharaki, Viiu Tuulik, Deniss Karai, Kristjan Pilt, Priit Roosipuu, and Tarmo Koppel for their interest and contribution to the research and experiments. I thank my family from the bottom of my heart for their support and help.

About the Author

Hiie Hinrikus was born in the Republic of Estonia in 1934. She received her primary and secondary education in Tallinn, Estonia. Afterward, she was educated in Physics, with a specialty in Radiophysics, at the Lomonosov Moscow State University. Her diploma work at the Sector of Radio Astronomy of the Lebedev Physical Institute of the Academy of Sciences of the USSR was related to the sensitivity of microwave radiometers. She completed her doctoral studies, investigating noise in masers at the USSR National Institute of Physical-Technical and Radio-Technical Measurements in the Moscow Region, and received a PhD in Radiophysics.

Hinrikus returned to Tallinn, Estonia, in 1968 and started work at the Department of Radio Engineering of the Tallinn Polytechnic Institute. She delivered lectures in Electrodynamics, Radio Wave Propagation, Antennas, and Microwave Engineering. Her research aimed at the application of lasers in communication and measurements, focusing on the noise in laser systems. Based on the results of these studies, she defended her thesis for the degree of Doctor of Sciences at the Institute of Radio and Electronic Engineering of the Academy of Sciences in Moscow. She was awarded the DSc degree in 1989.

After the Republic of Estonia restored its independence in 1991, Hinrikus started with research and education in biomedical engineering at the Tallinn University of Technology (TalTech). She initiated the establishment of the Center for Biomedical Engineering at the Tallinn University of Technology and the Estonian Society of Biomedical Engineering and Medical Physics in 1994. She was the first Chair of the BME Center and the first President of the Society. The BME Center has now been transferred to the Department of Health Technologies. The Society is responsible for accrediting Biomedical Engineers in Estonia.

Her research interests include the effects of microwaves on humans and their potential applications in medicine, with a focus on the sensitivity of the human brain to radiation. She is now Professor Emeritus but is still active in research on the microwave effects on the brain, which has been her interest for decades. She was awarded the Estonian National Science Prize in 2000, elected a fellow of the IFMBE International Academy for Medical and Biological Engineering in 2003 of the European Alliance for Medical and Biological Engineering and Science in 2013, and an IFMBE Honorary Life Member in 2018. She is a widow, a mother of a son, and a grandmother of three grandchildren, and lives in Tartu, Estonia.

About the Author

Symbols and Abbreviations

$\boldsymbol{B}$	the vector of magnetic flux density, magnetic induction (T)
$\boldsymbol{D}$	the vector of electric induction (electric displacement field) (C/m^2)
$\boldsymbol{D}$	the directivity of the antenna
$\boldsymbol{ds}$	surface element, vector
$\boldsymbol{E}$	the vector of electric field strength (V/m)
EEG	electroencephalography
$\boldsymbol{F}$	force (N)
$\boldsymbol{f}$	frequency (1/s)
$\boldsymbol{H}$	the vector of magnetic field strength (N/A^2)
$\boldsymbol{I}$	electric current (A)
IARC	International Agency for Research on Cancer
ICNIRP	International Commission on Non-Ionizing Radiation Protection
$\boldsymbol{j}$	the vector of electric current density (A/m)
$\boldsymbol{K}$	the rate of thermal reaction
MRI	magnetic resonance imaging
$\boldsymbol{n}$	unit vector
$\boldsymbol{P}$	power density (Pointing vector) (Wm^2)
$\boldsymbol{q}$	electric charge (C)
$\boldsymbol{r}$	distance (m)
$\boldsymbol{t}$	time (s)
$\boldsymbol{v}$	velocity (m/s)
$\boldsymbol{v_f}$	phase velocity (m/s)
$\boldsymbol{v_0}$	phase velocity in vacuum $v_0 = 3\times10^8$ (m/s)
$\boldsymbol{v_g}$	group velocity (m/s)
W	wave impedance
$\boldsymbol{w_E}$	Electric power density (W/m^3)
$\boldsymbol{w_H}$	Magnetic power density
∇	the vector of the three-dimensional gradient operator
$\boldsymbol{\varepsilon}$	electric permittivity (F/m)
$\boldsymbol{\varepsilon_0}$	electric permittivity of vacuum $\varepsilon_0 = (1/36\pi)10^{-9}$ (F/m)
$\boldsymbol{\varepsilon_c}$	relative permittivity (electric constant), $\varepsilon_c = \varepsilon/\varepsilon_0$
$\boldsymbol{\varphi}$	electric potential (V)
$\boldsymbol{\chi}$	electric susceptibility
$\boldsymbol{\lambda}$	wavelength (m)
$\boldsymbol{\lambda_0}$	wavelength in vacuum $\lambda_0 = 3 \times 10^8/f$ (m/s)
$\boldsymbol{\mu}$	magnetic permeability (H/m)
$\boldsymbol{\mu_0}$	magnetic permeability of vacuum $\mu_0 = 4\pi \times 10^{-7}$ (H/m)
$\boldsymbol{\sigma}$	electric conductivity (specific conductance) ($1/\Omega$ m)
$\boldsymbol{\omega}$	radial frequency $\omega = 2\pi f$ (2π/s)

Introduction

Humanity is adapted to living in natural electromagnetic radiation from the Sun. During the last century, the technical electromagnetic radiation has been added to the natural Sun's radiation. The properties of coherent technical electromagnetic radiation differ from the noncoherent natural radiation produced by the Sun. The natural electromagnetic radiation from the Sun is a random, noncoherent thermal radiation that is described only by its intensity. Technical radiation is a regular, periodic radiation described as a sinusoidal wave with a specific amplitude and phase. The technical electromagnetic radiation is a new physical environmental factor that may potentially affect people's physiology and health. People are concerned about the effects of microwave radiation on their health.

Over the last century, thousands of experimental studies have been performed by numerous independent research groups aimed at detecting biological effects of technical electromagnetic radiation. The results of these studies are diverse, but the majority of performed investigations have reported positive findings in various kinds of effects (Hinrikus et al., 2021; Lai and Levitt, 2022). Despite this, the existence of the effect of low-level radio radiation remains debated, and the mechanism underlying it remains unknown. This book aims to present a systematic and logical explanation for the mechanism of the nonthermal effects of low-level radiofrequency radiation, with a focus on microwave radiation on the human brain.

The principal factor limiting the possibility of low-level microwave radiation effects is thermal noise energy in biological systems, which is much higher than the energy of low-level radiation (Adair, 2003). However, due to the synchronous cumulative effect of coherent radiation on a huge number of dipolar molecules, low-level microwave radiation can overcome the thermal limit, and the effect is plausible (Hinrikus et al., 2018). There is no need to introduce a special new mechanism to describe the impact of low-level radiation. The dielectric polarization of tissues by the coherent microwave radiation makes the mechanism understandable.

This book presents the physical mechanism of the nonthermal effect of low-level microwave radiation. This book focuses on the aspects essential for understanding the nonthermal mechanism of low-level microwave effects involving relevant knowledge from electromagnetism, bioelectromagnetism, and nonlinear dynamics. This book provides a comprehensive description of various aspects of the nonthermal mechanism of low-level effects using original results published by the author's research group and reliable data from other peer-reviewed studies. The objective analysis of the mechanism and properties of low-level microwave effects should hopefully address concerns about its impact on health.

0.1 ELECTROMAGNETIC RADIATION

Electromagnetic radiation has a broad spectrum ranging from static fields at frequencies $f = 0$ and extremely low frequencies to gamma radiation at frequencies $f = 10^{25}$ and higher. Visible light forms a narrow part of the spectrum between the

TABLE 0.1
Spectrum of electromagnetic radiation

Frequency Hz	Frequency units	Wavelength	Band	
3×10	3 Hz	10,000 km	Ultra-long wave	Quasi-static
3×10^2	30 Hz	1000 km	Extremely low frequency (ELF)	electromagnetic field
3×10^3	300 Hz	100 km	Long wave	Low-frequency fields
3×10^4	30 kHz	10 km	Low-frequency (LF)	
3×10^5	300 kHz	1 km	Medium wave	Radiofrequency
3×10^6	3 MHz	100 m	Medium frequency (MF)	radiation
3×10^7	30 MHz	10 m	Shot wave High frequency (HF)	100 kHz–300 GHz
3×10^8	300 MHz	1 m	Ultrashort wave Ultra-high frequency (UHF)	
3×10^9	3 GHz	10 cm	Microwaves	
3×10^{10}	30 GHZ	1 cm	dm, cm, and mm waves	
3×10^{11}	300 GHz	1 mm		
3×10^{12}	3 THz	100 μm	Infrared	Optics
3×10^{13}	30 THz	10 μm	1 mm–750 nm	
3×10^{14}	300 THz	1 μm	Visible light 750–400 nm	
3×10^{15}	3 PHz	100 nm	Ultraviolet	Ionizing radiation
3×10^{16}	30 PHz	10 nm	400–10 nm	
3×10^{17}	300 PHz	1 nm	X-rays 10–0.01 nm	
$> 10^{18}$	3 EHz	100 pm	Gamma rays < 100 pm	

non-ionizing infrared and the ionizing ultraviolet radiation. The frequency structure of the electromagnetic spectrum is divided into traditional frequency bands (Table 0.1). Radiofrequency radiation covers the frequency band from 100 kHz to 300 GHz.

However, 100 kHz is not a frequency limit that separates low frequencies from radio frequencies in terms of their technical use. Ultra-low and low-frequency electromagnetic waves behave as waves on a global scale. Thus, ultra-long waves are successfully used in underwater communication and navigation. But long waves behave as quasi-static fields on a human scale because the wavelength is much longer than the dimensions of the object. The physics underlying the possible effects of the radiation differs in the cases of quasi-static fields and radiofrequency radiation. Therefore, the potential physiological and biological impacts of the quasi-static fields and radiation are different and should be discussed separately.

Electrical charges and currents are the sources of electromagnetic fields. Radiofrequency radiation is generated by the acceleration of the fluxes of free charged particles (electrons) or their movement in a magnetic field. Infrared radiation is produced by the vibrations of molecules. Visible and ultraviolet light is generated by electronic transitions in chemical reactions and the process of ionization of atoms and molecules. X-rays are generated by the acceleration of electrons or by the ionization of atoms. Gamma rays are produced by high-energy nuclear processes. The energies of quanta of X-rays and gamma rays are sufficient to break the structure of molecules and atoms, and they are dangerous for living organisms.

The Sun is the largest and dominant source of electromagnetic radiation. The impact of other stars is minimal. The natural electromagnetic radiation is thermal radiation generated by the Sun. The Sun can be considered a blackbody at a temperature of about 5770 K.

0.2 STRUCTURE OF THIS BOOK

Electromagnetic radiation is a physical factor. Knowledge about electromagnetism is the basis for understanding the effect and describing its properties. The first part of this book focuses on the general principles of electromagnetism underlying the aspects essential to understanding the low-level effects.

The first chapter includes the most general rules applicable to the whole spectrum of electromagnetic radiation. Four linear Maxwell's equations and three material equations that relate fields to the medium are the basis of electromagnetism, comprehensively describing all electromagnetic processes.

Polarization is the parameter that determines the relationship between the electric field and the medium. Dielectric polarization describes the primary effect of the electric field in a medium. Dielectric polarization depends linearly on electric field strength. The nonthermal effect of the electromagnetic field is expected to be related to the electric field strength rather than the energy of the field.

Maxwell's equations describe the relationship between electric and magnetic fields and the relationship between fields and their sources, charges, and currents. Due to the linearity of Maxwell's equations, the principle of superposition is valid for electromagnetic fields in a linear medium. The total field is the sum of all fields that exist in a medium.

The second chapter presents the general theory and properties of electromagnetic waves in the radiofrequency band. The wave equation derived from Maxwell's equations is linear. The parameters and properties of the wave in a medium depend on the electric permittivity of the medium. Inhomogeneity of the medium causes refraction and reflection of the wave. Reflection leads to the formation of a standing wave characterized by the reactive fields, appearance of maximums and minimums in the wave amplitude, and inhomogeneous distribution of power density. The direction of wave propagation and polarization of the electric field can be changed upon reflection, but not its coherence. The interference between partial waves from different parts of a large antenna results in a nonhomogeneous spatial distribution of the radiation, which determines the directivity of the antenna.

The third chapter introduces the electrical parameters of human tissues, with a focus on microwave frequencies. Human tissues can be considered a lossy dielectric at microwave frequencies. Two electric parameters, electric permittivity and conductivity, determine the electrical properties of tissues. Dielectric permittivity describes dielectric polarization, the primary effect of the external radiofrequency radiation in tissues caused by the radiation. The electric properties of tissues depend on temperature and frequency but not on the strength of the electric field. Dielectric polarization is a linear process. The high content of water in the biological tissues is important in determining the electrical properties of tissues. The electric permittivity (related to low-level nonthermal effects) decreases, and the conductivity (related to thermal effects) increases with frequency. The parameters of electromagnetic waves in biological tissues differ from those in air. About half of the radiation power is reflected from the skin.

The fourth chapter describes the principles of bioelectromagnetism and the behaviors of nonlinear biological systems. The neuronal functioning of the brain comprises the complex cooperation of 10^{11} neurons connected by neural fibers and synapses (Malmivuo and Plonsey, 1995). Brain neuronal activity has an oscillating nature. The propagation of nerve pulses between neurons determines the spectrum of brain bioelectric signals from very low frequencies to approximately 2000 Hz, the spiking frequency of a single neuron. The brain's bioelectrical oscillations form an EEG signal, which reflects the brain's functioning. Biological systems are nonlinear (Adey and Lawrence, 1984; Barnes and Hu, 1980). The dynamics of a neuronal system are described by the nonlinear Hodgkin–Huxley model. The dynamics of a nonlinear living system differ from those of a linear non-living system described by the linear Maxwell's equations. The complexity and nonlinearity of living systems make the results of the impact of external electromagnetic radiation chaotic and unpredictable.

The second part of this book is based mainly on the generalization of original results published by the author's research group relevant to understanding the physical mechanisms of low-level radiation and its impact on the brain neuronal activity.

The fifth chapter presents a logical structure of the physical mechanism of microwave radiation. Coherence of the radiation is crucial in nonthermal mechanisms of low-level radiation (Hinrikus et al., 2018). The cumulative effect of minor disturbances enables the coherent microwave radiation to overcome the thermal noise limit kT. The origin of the effects is the dielectric polarization of biological tissues caused by external electric fields. The dielectric polarization disturbs hydrogen bonds, leading to increased diffusion and oxidative stress. Microwave radiation can affect the majority of physiological processes due to its impact on diffusion (Hinrikus et al., 2015; Hinrikus et al., 2017; Hinrikus et al., 2018). The increased oxidative stress, related to radiation, can have a broad spectrum of biological consequences. The modulation is crucial in revealing the effect, and the features of the effect depend on the modulation frequency (Hinrikus et al., 2008a, 2008b). The modulated low-level microwave radiation can cause parametric excitation of brain bioelectric oscillations at predetermined frequencies (Hinrikus et al., 2011b). The nonthermal mechanism has no physical threshold (Hinrikus et al., 2021). All steps of the mechanism are supported by experiments carried out by several independent researchers. This chapter proposes a novel approach and demonstrates that, rather than a special mechanism, dielectric polarization, a well-known and commonly accepted phenomenon, can trigger small initial changes sufficient to result in significant alterations in nonlinear living systems and cause nonthermal effects.

The sixth chapter focuses on the analysis of experimental results of low-level microwave effects on the human brain. An analytical review of published experimental results shows that the majority of studies by many independent authors on microwave effects on the human brain have reported positive results. Still, the features of the effects are masked due to the diversity of methods and the high level of natural variability of the EEG signal (Hinrikus et al., 2021). The developed advanced method of integration of differences provides a way to determine the features of the microwave effect on the human brain (Hinrikus et al., 2007). The results reported in several publications by the author's research group show that the effect depends on modulation frequency and radiation level but much less on the carrier frequency (Hinrikus et al., 2008a, 2008b, 2021, 2022; Suhhova et al., 2013). The effect is proportional to electric field strength rather than SAR (Suhhova et al., 2013). The effect differs between individuals. Only a minority, less than 30% of exposed participants, were significantly affected (Hinrikus et al., 2008b). The differences are more evident in the EEG signal, which directly reflects brain neuronal activity, rather than in behavior or sleep quality (Lass et al., 2002; Rodina et al., 2005; Hinrikus et al., 2021).

The third part of this book presents a discussion on biological and health effects, as well as health protection limits.

The seventh chapter discusses the biological effects of microwave radiation and its possible impact on health, considering the published reviews and the knowledge about the features of biological processes. The oxidative stress caused by microwave radiation is known to be related to many different diseases. Therefore, it is not surprising that a broad range of different effects in various tissues and cells have been reported in a large number of studies by many research teams (Lai and Levitt, 2022). Due to the body's ability to repair changes, not all biological effects cause an impact on health. The transformation of the biological effects of radiation into health effects is a nonlinear, chaotic process with unpredictable outcomes. The impact of microwave radiation on humans can be evaluated only using epidemiologic methods. However, today, it is not possible to find a reference group living in a radiation-free environment. Human health is affected by various factors, making it complicated to differentiate the impact of one of these, electromagnetic radiation. A long latency period of cancer, about eight years, makes it difficult to determine the initial cause of cancer in a nonlinear living system. The results of studies in laboratory animals have demonstrated an increase in cancer cases in exposed animals compared to sham (Falcioni et al., 2018; NTP, 2018). Cellular studies demonstrate biological effects, rather than direct effects on health. The health effects of microwave radiation are still under discussion, and the high variability between results suggests the need for further high-quality studies. The thermal microwave auditory effect is the only one described theoretically (Lin, 2022). The ability of microwave radiation to produce Havana syndrome is hardly possible at the current level of technology. Thermal microwave radiation is in clinical use for electrotherapy and hyperthermia. Low-level microwave radiation is used in microwave imaging, and promising results have been reported in brain stimulation (Camacho-Conde et al., 2022; Carr, 1989; Hinrikus et al., 1998; Hinrikus and Riipulk, 2006).

The eighth chapter aims to discuss the current situation with radiofrequency radiation in the living environment worldwide and the criteria for health protection limits, considering both thermal and nonthermal mechanisms. The chapter combines information about existing limits and published data about the current levels of radiofrequency radiation in different countries to analyze various criteria for updating the limits. The countries dominantly follow the ICNIRP guidelines (Ramirez-Vazquez et al., 2024). However, the current guidelines are based on short-term thermal effects and do not consider long-term low-level health effects (Barnes and Greenebaum, 2020; Hinrikus et al., 2023). Nowadays, the measured levels of radiofrequency radiation worldwide are limited to 2 V/m, much lower than recommendations for health protection in the ICNIRP guidelines, 61 V/m. The existing guidelines, which are based on thermal effects, should be revised to consider not only thermal but also low-level nonthermal effects and long-term exposure. Electric field strength is an appropriate parameter for describing low-level radiofrequency radiation effects. Since there is no physical threshold for nonthermal effects, health protection limits can be justified considering pragmatic aspects (Hinrikus et al., 2023). The level of radiofrequency radiation in the era before mobile telecommunication, 6 V/m, can be considered a recommended limit of health protection. The stricter restrictions do not prevent the development of technology. The total level of radiofrequency radiation decreases with the development of telecommunication technology (Hinrikus et al., 2023). Due to the chaotic nature of health effects, it is most reasonable to follow a precautionary principle and the ALARA approach as low as reasonably achievable to minimize the health risks related to radiofrequency radiation.

0.3 CONCLUSIONS

This is the first book presenting a systematic description of the physical mechanism and characteristics of low-level, microwave radiation nonthermal effects. This book introduces a new concept of the nonthermal mechanism of microwave radiation effect, focusing on the aspects essential for understanding the nonthermal mechanism, considering the nonlinearity of neurophysiological and biological processes. The incorporation of relevant knowledge from electromagnetism, bioelectromagnetism, and nonlinear dynamics enables a systematic, logical description of all steps in the nonthermal mechanism. This book contributes to understanding the low-level microwave effects in biological systems and possible related health problems. The introduced novel physical model of microwave radiation nonthermal effect makes it possible to describe the nature of the effects.

- The cumulative effect of minor disturbances caused by coherent microwave radiation in tissues makes it possible to overcome the thermal noise limit kT. The impact of low-level microwave radiation on biological processes is likely.
- The dielectric polarization, rather than a special mechanism, can trigger small initial changes sufficient to result in significant alterations in nonlinear living systems, causing nonthermal effects.
- The predominantly positive results from a broad range of different biological effects in multiple tissues and cells reported in a large number of low-level microwave studies performed by numerous independent researchers support the capability of low-level radiation to produce biological effects.
- The modulation is crucial in revealing the effect, and the features of the effect depend on the modulation frequency.

- The effect on the brain differs between individuals. Only a minor part of people are significantly affected.
- Due to the body's ability to repair changes, not all biological effects cause an impact on health. The transformation of the biological effects of radiation into health effects is a nonlinear, chaotic process with unpredictable outcomes.

REFERENCES

Adey WR, Lawrence AF, (1984). *Nonlinear electrodynamics in biological systems.* Plenum Press, New York.

Barnes FS, Hu C-LJ, (1980). Nonlinear interactions of electromagnetic waves with biological materials. In: PLE Uslenghi, editor. *Nonlinear electromagnetics.* Academic Press, New York, p. 391.

Barnes F, Greenebaum B, (2020). Setting guidelines for electromagnetic levels and research needs. *Bioelectromagnetics.* 41(5):392–397.

Camacho-Conde JA, Gonzalez-Bermudez MDR, Carretero-Rey M, Khan ZU, (2022). Brain stimulation: a therapeutic approach for the treatment of neurological disorders. *CNS Neurosci Ther.* 28(1):5–18. doi: 10.1111/cns.13769.

Carr, K, (1989). Microwave radiometry: its importance to the detection of cancer. *IEEE Trans. Microwave Theory Tech.* MTT37:1862–1869.

Falcioni L, Bua L, Tibaldi E, Lauriola M, De Angelis L, Gnudi F, Mandrioli D, Manservigi M, Manservisi F, Manzoli I, Menghetti I, Montella R, Panzacchi S, Sgargi D, Strollo V, Vornoli A, Belpoggi F, (2018). Report of final results regarding brain and heart tumors in Sprague-Dawley rats exposed from prenatal life until natural death to mobile phone radiofrequency field representative of a 1.8 GHz GSM base station environmental emission. *Environ Res.*165:496–503. doi: 10.1016/j.envres.2018.01.037.

Hinrikus H, Riipulk J, Neemela T, Podra H, (1998). Sensitivity of microwave radiometer for tissue screening. *Proc Estonian Acad Sci Eng.* 4:165–177.

Hinrikus H, Riipulk J, (2006). Microwave imaging. In M Akay, editor. *Wiley encyclopedia of biomedical engineering.* Wiley, New Jersey, pp. 2329–2340.

Hinrikus H, Bachmann M, Kalda J, Sakki M, Lass J, Tomson R, (2007). Methods of electroencephalographic signal analysis for detection of small hidden changes. *Nonlinear Biomed Phys.* 1(1):9. doi: 10.1186/1753-4631-1-9.

Hinrikus H, Bachmann M, Lass J, Tomson R, Tuulik V, (2008a). Effect of 7, 14 and 21 Hz modulated 450 MHz microwave radiation on human electroencephalographic rhythms. *Int. J. Rad. Biol.* 84:69–79.

Hinrikus H, Bachmann M, Lass J, Karai D, Tuulik V, (2008b). Effect of low frequency modulated microwave exposure on human EEG: individual sensitivity. *Bioelectromagnetics.* 29(7):527–538.

Hinrikus H, Bachmann M, Karai D, Klonowski W, Lass J, Stepien P, Stepien R, Tuulik V, (2011a). Higuchi's fractal dimension for analysis of the effect of external periodic stressor on electrical oscillations in the brain. *Med Biol Eng Comput.* 49(5):585–591.

Hinrikus H, Bachmann M, Lass J. (2011b). Parametric mechanism of excitation of the electroencephalographic rhythms by modulated microwave radiation. *Int J Radiat Biol.* 87(11):1077–1085.

Hinrikus H, Lass J, Karai D, Pilt K, Bachmann M, (2015). Microwave effect on diffusion: a possible mechanism for nonthermal effect. *Electromagn Biol Med.* 34(4):327–333.

Hinrikus H, Bachmann M, Karai D, Lass J, (2017). Mechanism of low-level microwave radiation effect on the nervous system. *Electromagn Biol Med.* 36(2):202–212.

Hinrikus H, Bachmann M, Lass J, (2018). Understanding the physical mechanism of low-level microwave radiation effect. *Int J Radiat Biol.* 94(10):877–882.

Hinrikus H, Lass J, Bachmann, (2021).Threshold of radiofrequency electromagnetic field effect on human brain. *Int J Radiat Biol.* 97(11):1505–1515.

Hinrikus H, Koppel T, Lass J, Orru H, Roosipuu P, Bachmann M, (2022). Possible health effects on the human brain by various generations of mobile telecommunication: a review based estimation of 5G impact. *Int J Radiat Biol.* 98(7):1210–1221.

Hinrikus H, Koppel T, Lass J, Roosipuu P, Bachmann M, (2023). Limiting exposure to radiofrequency radiation: the principles and possible criteria for health protection. *Int J Radiat Biol.* 99(8):1167–1177.

Lai H, Levitt BB. (2022). The roles of intensity, exposure duration, and modulation on the biological effects of radiofrequency radiation and exposure guidelines. *Electromagn Biol Med.* 41(2):230–255.

Lass J, Tuulik V, Ferenets R, Riisalo R, Hinrikus H, (2002). Effects of 7 Hz-modulated 450 MHz electromagnetic radiation on human performance in visual memory tasks. *Int J Radiat Biol.* 78(10):937–944.

Lin, JC. (2022). The Microwave auditory effect. *IEEE J Electromagn RF Microw Med Biol.* 6(1):16–28.

Malmivuo J, Plonsey R, (1995). *Bioelectromagnetism.* Oxford University Press, New York.

NTP, (2018). National Toxicology Program. Toxicology and carcinogenesis studies in Sprague Dawley (Hsd: Sprague Dawley SD) rats exposed to whole-body radio frequency radiation at a frequency (900 MHz) and modulations (GSM and CDMA) used by cell phones. *Natl Toxicol Program Tech Rep Ser.* 2018 Nov;(595): NTP-TR-595. doi: 10.22427/NTP-TR-595.

Ramirez-Vazquez R, Escobar I, Vandenbosch GAE, Arribas E, (2024). Personal exposure to radiofrequency electromagnetic fields: a comparative analysis of international, national, and regional guidelines. *Environ Res.* Apr 1;246:118124. doi: 10.1016/j.envres.2024.118124.

Rodina A, Lass J, Riipulk J, Bachmann T, Hinrikus H, (2005). Study of effects of low level microwave field by method of face masking. *Bioelectromagnetics.* 26(7):571–577.

Suhhova A, Bachmann M, Karai D, Lass J, Hinrikus H, (2013). Effect of microwave radiation on human EEG at two different levels of exposure. *Bioelectromagnetics.* 34(4):264–274.

Part I

Basic knowledge about electromagnetism and bioelectromagnetism

Knowledge is experience. Everything else is just information.

Albert Einstein

DOI: 10.1201/9781003729976-1

1 Electromagnetic fields
Fundamental laws of electromagnetism and properties of the fields

HIGHLIGHTS

- The impact of an electromagnetic field on a charge located in the field is unavoidable.
- Polarization is the parameter that determines the relationship between the electric field and the medium. Dielectric polarization describes the primary effect of the electric field in a medium. Dielectric polarization depends linearly on electric field strength.
- The non-thermal effect of the electromagnetic field is expected to be related to the electric field strength rather than energy.
- Due to the linearity of Maxwell's equations, the principle of superposition is valid for electromagnetic fields in a linear medium, and the total field is a sum of all fields existing in a medium.
- Electric field is always perpendicular and magnetic field parallel to conductive objects.

1.1 INTRODUCTION

What is a field? This simple question is not easy to answer. A field can be understood only through its effect on certain specific objects that are subject to its influence. A field can be understood as a part of the space where conditions exist for such an effect to occur. The gravitational field is determined by a force acting on a mass located in the field. The field is generated by another mass. The electric field is a force acting on a charge located in the field. The field, in turn, is generated by another charge. The force on a certain specific object, mass or charge, is unavoidable if the object is located in the field. Electric forces keep together atoms, molecules, and cells. The functioning of living nature is based on the cooperation of charged particles. The impact of an external electromagnetic field on the charges is unavoidable. The basic formulas of electromagnetism describe the direct effect of external electromagnetic fields on the charges in living structures.

The theory of electromagnetic fields and waves has been presented in the classical book by Ramo and Whinnery (Ramo and Whinnery, 1944) at a level suitable for readers with a background in the general course of physics. The third author, van

DOI: 10.1201/9781003729976-2

Duzer, has been added later (Ramo et al., 1965, 1994). The book emphasizes physical understanding and has proved its value through several editions over the decades. The course on electromagnetic field theory by Zahn (Zahn, 1979) is a good presentation of the theory, supported by solving problems and related calculations.

This chapter underlines the main rules of electromagnetism important for understanding the mechanism of electromagnetic field effects. Six axioms, four Maxwell's equations, and three material equations comprise the basis of the theory of electromagnetic fields. The axioms are formulated as a theoretical generalization of experimental results that apply across the entire electromagnetic spectrum.

1.2 SOURCES OF ELECTROMAGNETIC FIELD

The concepts of the electric and magnetic fields can be induced based on electric charges. An electric field is generated by and related to electric charges. An electric field creates a force that affects the electric charge located in it. According to Coulomb's law, force $\boldsymbol{F}$ between two charges q_1 and q_2

$$\boldsymbol{F} = \frac{q_1 q_2}{4\pi\varepsilon r^2}\boldsymbol{n}, \tag{1.1}$$

where ε is electric permittivity, r is the distance between the charges, and $\boldsymbol{n}$ is a unit vector directed from one charge to another (Figure 1.1A). In a vacuum, the electric permittivity $\varepsilon_0 = \frac{1}{36\pi}10^{-9}\,\frac{\mathrm{F}}{\mathrm{m}}$. The electric properties of a medium are described using relative electric permittivity (electric constant) $\varepsilon_r = \frac{\varepsilon}{\varepsilon_0}$.

Electric field strength is the force acting on a unit charge. The vector of electric field $\boldsymbol{E}$ can be defined from the formula 1.1 as

$$\boldsymbol{E} = \frac{q_1}{4\pi\varepsilon r^2}\boldsymbol{n} \tag{1.2}$$

The electric Lorenz force (Figure 1.1B) is proportional to the electric field $\boldsymbol{E}$ and the quantity of the charge q

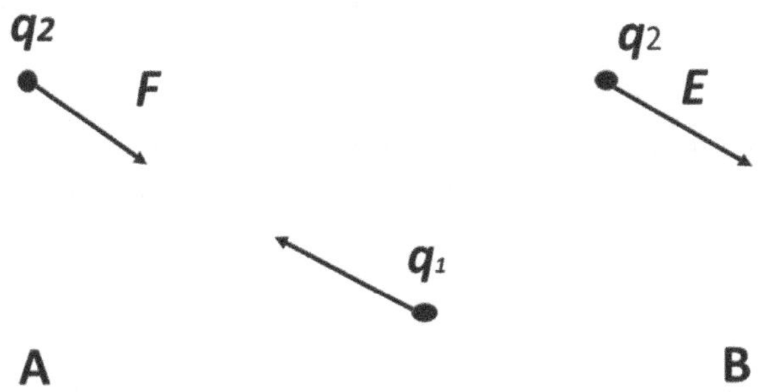

FIGURE 1.1 (A) Coulomb's law. (B) Electric Lorentz force.

$$F = qE \tag{1.3}$$

The electric Lorenz force produced by an electric field affects each charge located in the electric field and does not depend on the other fields. In a linear medium, where several fields generated by different independent sources exist simultaneously, the Lorenz forces produced by the different fields are independent.

A charge generates an electric force around it in all directions. Gauss's law states that the total electric flux through a closed surface S is determined by the charge in the volume surrounded by the surface:

$$\int_s \mathbf{D}d\mathbf{s} = q \tag{1.4}$$

where vector $\mathbf{D}$ is the electric displacement field or electric induction. The electric induction is proportional to the electric field strength and electric permittivity

$$\mathbf{D} = \varepsilon \mathbf{E} \tag{1.5}$$

There are no free magnetic charges in nature, and the magnetic field does not affect a stationary electric charge. The magnetic field can be produced by moving electric charges and conduction current (Figure 1.2). Ampère's law determines the relationship between the circulation of the magnetic field around the closed loop path and the total electric current I passing through the loop:

$$\int_l \mathbf{B}d\mathbf{l} = \mu I \tag{1.6}$$

where the vector $\mathbf{B}$ is magnetic flux density or magnetic induction. Magnetic induction $\mathbf{B}$ is proportional to magnetic field $\mathbf{H}$ and magnetic permeability μ

$$\mathbf{B} = \mu \mathbf{H} \tag{1.7}$$

In vacuum, $\mu_0 = 4\pi 10^{-7}\left[\frac{H}{m}\right]$. The magnetic properties of a medium are described using relative magnetic permeability $\mu_r = \frac{\mu}{\mu_0}$.

The magnetic field doesn't affect stationary but only moving electric charges. A magnetic Lorenz force $\mathbf{F}_m$ affects a charge q moving with the velocity $\mathbf{v}$ in a magnetic field,

$$\mathbf{F}_m = q\mu \mathbf{v} \times \mathbf{H} \tag{1.8}$$

Magnetic force $\mathbf{F}_m$ in 1.4 is perpendicular to the magnetic field $\mathbf{H}$ and the velocity of charge movement $\mathbf{v}$ (Figure 1.2B).

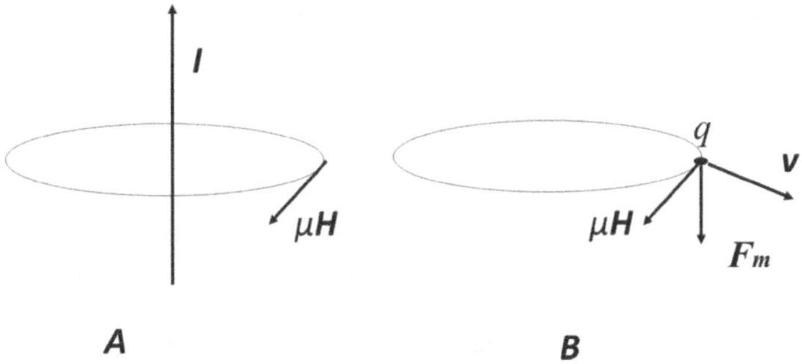

FIGURE 1.2 (A) Ampère's law. (B) Magnetic Lorenz force.

Materials can be diamagnetic where $\mu_r = 1$, paramagnetic where magnetic permeability $\mu_r > 1$, and ferromagnetic, where magnetic permeability $\mu_r >> 1$. There are no unipolar magnetic charges in nature. Therefore, the majority of materials are diamagnetic. Paramagnetism can be related to electron spins in some molecules like oxygen or some crystals (corundum). Ferromagnetic properties are characteristic of some metals like iron.

In a conductive medium, the electric Lorenz's force affects free charges and the moving charges generate conductive current in a medium. In a generalized Ohm's law, in which voltage and current are replaced by electric field strength and current density, the density of conduction current

$$j = \sigma E \tag{1.9}$$

where σ is the specific conductance of the medium.

The Lorenz's forces affect each charge in the electromagnetic field independently. If there are fields created by several sources simultaneously in a linear medium, the effect of each field on the charges located there is independent.

Three equations, 1.5, 1.7, and 1.8, describe the relationship of electric and magnetic fields with a medium. These equations are axioms that consist part of the basis of the general theory of electromagnetism.

1.3 DIELECTRIC POLARIZATION

A medium contains bound and free charges. The movement of free charges results in the conduction current (Eq. 1.9). The charges bound in the structure of a dielectric nonconductive material cannot move freely but only be displaced. The charges are distributed in the material in such a way that the strength of the electric field generated by them is neutral, equal to zero. Otherwise, the electric forces displace the charges to restore the neutral situation. An external electric field affects the charges, causes their displacement, and disturbs the neutral state of the material. In response to the external electric field, the charges inside the medium should be displaced in such a way that the electric field in the medium restores its zero value. The displacement of electric charges causes the polarization of a dielectric medium. The vector of polarization in a linear medium

$$P = \chi E \tag{1.10}$$

where χ is electric susceptibility.

The polarization is the parameter that determines the relationship between the electric field and media. Dielectric polarization produces the primary effect of the electric field in media.

The vector of the electric displacement field inside a material

$$D = \varepsilon_0 E + \varepsilon_0 \chi E = \varepsilon E \tag{1.11}$$

The electric polarization can be caused by different mechanisms (King and Smith, 1981). The displacement of negative and positive charges in the structure of a material, like electronic clouds in an atom regarding the nucleus, results in electronic polarization. The displacement of ions inside molecules results in ionic polarization (Figure 1.3a). The rotation and reorganization of dipolar molecules result in orientational polarization (Figure 1.3b).

The electronic and ionic polarizations are weaker compared to rotational polarization. The charged ions inside molecules are connected by very strong intermolecular electric forces. The dipolar molecules are connected by much weaker intramolecular electric forces.

The external electric field is very weak compared to both, the intermolecular and intramolecular electric forces. However, due to the large number of atoms, 6.02×10^{23} per gram molecular weight (Avogadro's number), the light imbalance in the distribution of charges can have a remarkable cumulative effect on the electric field inside the material.

The orientational polarization dominates at radiofrequencies. Water molecules are electric dipoles, and the electric constant of water is high, close to 80 at radiofrequencies. The ionic polarization becomes dominant at frequencies higher than about

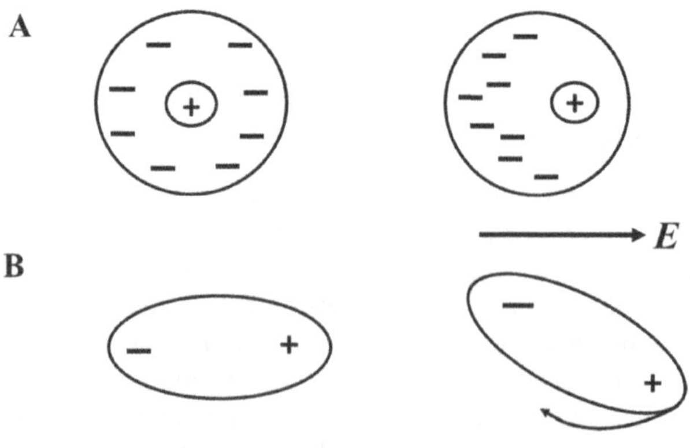

FIGURE 1.3 (A) Displacement polarization. (B) Orientational polarization.

200 GHz, and electronic polarization in optical frequencies where the electric constant is about 2. The frequency dependence of the polarization depends on the inertia of particles, relaxation, and resonance processes in the medium.

1.4 SCALAR FIELDS AND VECTOR FIELDS

Let's look at some mathematical concepts from general field theory that are in use in describing an electromagnetic field. The spatial properties of a field in three-dimensional space are described by a vector

$$\nabla = \frac{\partial}{\partial x}\boldsymbol{n}_x + \frac{\partial}{\partial y}\boldsymbol{n}_y + \frac{\partial}{\partial z}\boldsymbol{n}_z \tag{1.12}$$

where $\boldsymbol{n}_x$, $\boldsymbol{n}_y$, and $\boldsymbol{n}_z$ are unit vectors toward the coordinates.

A scalar field is a part of space, each point of which is characterized by the value of a certain scalar quantity (fields of temperature, density, potential, etc.). In the scalar field, there are so-called equipotential surfaces on which the quantity φ has a constant value (Figure 1.4).

The spatial vector derivation of a scalar field is a gradient that describes the change of the potential. In three-dimensional space, the gradient is expressed as a vector operator.

The gradient of a scalar field is a vector aimed from its lower values to higher values. The electric potential φ, the potential, is a scalar quantity. The field of electric potential (Figure 1.3) is a scalar field described by its gradient.

The gradient is the vector that describes the spatial change of a potential field. The absolute value of the electric potential does not affect the gradient. The electric field strength is a vector directed from a higher potential to a lower potential

$$\boldsymbol{E} = -\nabla\varphi \tag{1.13}$$

A **vector field** is a part of space, each point of which is characterized by the direction and value of a certain vector quantity (electric field, magnetic field, gravitation

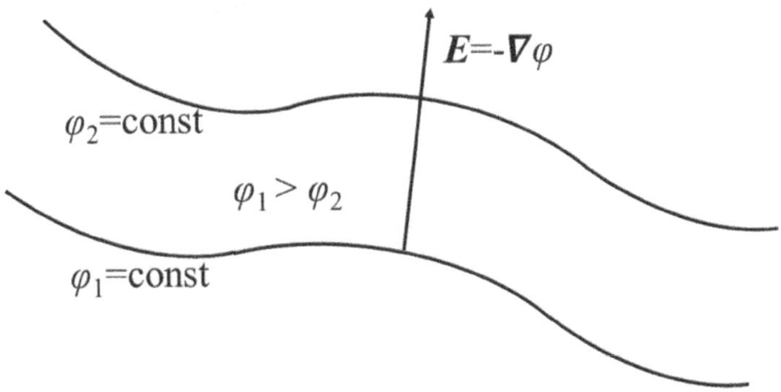

FIGURE 1.4 Gradient of the electric potential field.

field, etc.). The lines to which the field vectors are tangent at each point are the field lines.

Field lines can have very different structures. Two types of field lines can be distinguished: divergent and vortices.

The scalar (dot) product ∇ of the vector field denotes divergence. The divergence of a vector field is a scalar quantity that describes the source of the vector field. The divergence of the field or dispersion at a given point is the limit reached by the vector flow through a closed surface S in the space bounded by the same surface. The vector flow through a surface is related only to its component perpendicular to the surface.

The positive value of divergence div$\boldsymbol{D} > 0$ means that a given point is the source of the field (Figure 1.5). The negative value of divergence means that at the given point the field is absorbed. The zero divergence means that the field has no sources.

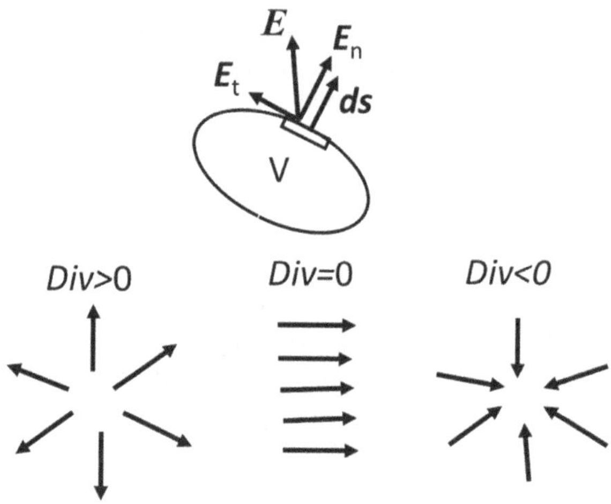

FIGURE 1.5 Divergence of a vector field.

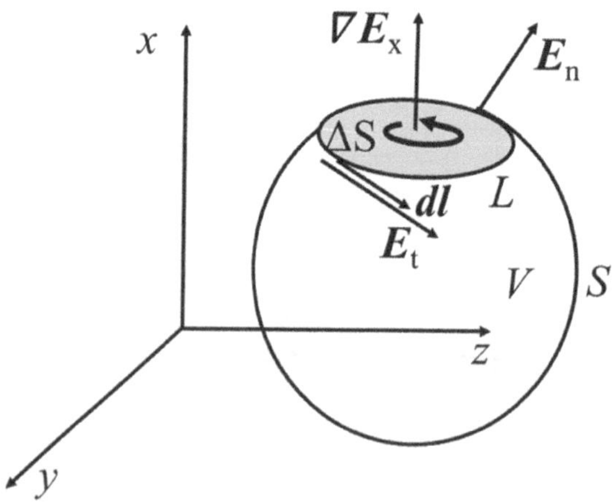

FIGURE 1.6 The curl x component of a vector field.

Vector(cross) product ∇× denotes curl. The curl is a vector that describes the vortices in the vector field. Curl is related to the vector's component parallel to the surface.

A curl is a vector; its x component is an elementary vortex. It can be represented as the limit of circulation along the borderline L of the elemental surface S when $S \to 0$ (Figure 1.6).

The identities of vector analysis are beneficial in describing the properties of electromagnetic fields.

- The divergence of the curl is zero, $\nabla \cdot (\nabla \times \boldsymbol{E}) = 0$. It is determined by the fact that the divergence is related to normal, but the vortex to tangent components of the vector field. There are no sources in the vorticity field.
- The gradient field has no vortices, $\nabla \times (\nabla \boldsymbol{A}) = 0$. The field of electric potential has no vortices.

1.5 ELECTROMAGNETIC FIELD EQUATIONS

The equations above are based on experimental results; the electric and magnetic fields have been induced through their connectivity with electric charges. Maxwell's equations introduce more general insight into the electromagnetic fields, including not only the connectivity of the fields with charges but also the mutual connection of electric and magnetic fields to each other. James Clerk Maxwell published the four equations first in 1865 and "A Treatise on Electricity and Magnetism" in 1873. The equations are based on laws previously formulated and experimentally proven by other scientists: Ampère's law, Faraday's law, Gauss's electric field law, and Gauss's magnetic field law. Maxwell reformulated Faraday's law and added a crucial correction to Ampère's law, displacement current. Maxwell's crucial contribution was that he combined these laws into a system of equations that completely describes the behaviors of electromagnetic fields. These equations have been the basis of the electromagnetic theory for centuries. The university courses in electromagnetic fields and waves, from classical ones by Ramo and Whinnery to more modern ones by Zahn, are based on Maxwell's equations.

Today, Maxwell's equations are most frequently used in their differential form:

$$\nabla \times \boldsymbol{H} = \varepsilon \frac{\partial \boldsymbol{E}}{\partial t} + \boldsymbol{j} \tag{1.12}$$

$$\nabla \times \boldsymbol{E} = - \frac{\partial \boldsymbol{H}}{\partial t} \tag{1.13}$$

$$\nabla \cdot \varepsilon \boldsymbol{E} = \rho \tag{1.14}$$

$$\nabla \cdot \mu \boldsymbol{H} = 0 \tag{1.15}$$

The first Maxwell's equation (Eq. 1.12) is generalized Ampère's law. Maxwell extended the law initially formulated for the conduction current existing only in a conductive medium to a dielectric medium, introducing displacement current as ***dD***/*dt*. The first Maxwell's equation switches the vortices of the magnetic field with electric displacement and conduction currents.

The second Maxwell's equation (Eq. 1.13) is a generalized Faraday's law and switches the vortices of the electric field with an alternating magnetic field. The third equation (Eq. 1.14) is a generalized Gauss's law and confirms that electric charge is a source of electric flux. The fourth equation (Eq. 1.15) confirms that there are no unipolar magnetic charges in nature and states that the magnetic field has no sources.

Four Maxwell's equations (Eqs. 1.12–1.15) and three material equations (Eqs. 1.5, 1.7, and 1.9), which describe the relationship of field vectors with media, present a mathematical description of seven experimentally proved axioms. These seven axioms consist of all the knowledge that is sufficient and exhaustive to describe and understand all electromagnetic processes and phenomena. From these axioms, the main properties of the electromagnetic field can be deduced.

1. The first (Eq. 1.12) and second (Eq. 1.13) Maxwell's equations state that the vectors of electric field $\boldsymbol{E}$ and magnetic field $\boldsymbol{H}$ are perpendicular to each other.
2. The second (Eq. 1.13) and third (Eq. 1.14) Maxwell's equations state that the electric field can be generated by an alternating magnetic field or by an electric charge.
3. The first Maxwell's equation (Eq. 1.12) states that the magnetic field can be generated by the alternating electric field or electric current.
4. Alternating electric and magnetic fields are connected and cannot exist without each other (Eqs. 1.12 and 1.13). Only static $\left(\frac{\partial}{\partial t}=0\right)$ or quasi-static $\left(\frac{\partial}{\partial t}\ll 0\right)$ electric and magnetic fields related to constant charge (Eq. 1.14) or direct current (Eq. 1.12) can exist independently of each other.
5. First (Eq. 1.12) and second (Eq. 1.13) Maxwell's equations show that alternating electric and magnetic fields $\left(\frac{\partial}{\partial x}\neq 0\right)$ can exist, once generated, independently of their sources, charge, or electric conduction current.
6. Maxwell's equations include only linear operators. In a linear medium, a linear combination of solutions is also linear. Electromagnetic processes in a linear medium are linear.
7. The principle of superposition is valid for electromagnetic fields in a linear medium. Separate electromagnetic fields are independent and not affected by other electromagnetic fields in a linear medium. The interference of coherent electromagnetic fields results from the principle of superposition.

1.6 BOUNDARY CONDITIONS

Maxwell's equations contain spatial derivatives of the field and induction vectors. For these derivatives to be determined, both the field vectors and the environmental parameters must be continuous functions of the coordinates. In reality, one often encounters cases where the requirement of continuity of environmental parameters is violated. Moreover, such situations occur frequently and are of great importance in electromagnetism. An example is metal surfaces in contact with air or another dielectric or skin with air. Maxwell's field equations cannot be directly used in such cases. However, the theory must allow solving tasks for calculating fields even in cases where the parameters of the environment change rapidly. Consequently, additional conditions are needed that would link the field and induction vectors at the boundary between the two environments. Such conditions are called boundary conditions.

Like all other laws of the electromagnetic field, the boundary conditions are derived from Maxwell's equations. A special technique is used to overcome a contradiction that these equations don't work when the parameters of the environment change suddenly. First, it is assumed that the boundary between the two media has a finite thickness within which the change of the parameters is continuous. Next, Maxwell's equations are used in the boundary zone and the thickness of the zone is approximated to zero. The derived boundary conditions for electric and magnetic vectors are similar, but the relations between the normal and parallel-to-the-boundary-surface components of the field and induction vectors are different.

On the boundary between the media without surface charges (dielectric, medium with relatively low electrical conductivity), the normal components of the induction vectors in the first and second medium D_{1n} and D_{2n}, $B_{1n,}$ and B_{2n} are continuous, but the normal components of the electric and magnetic field change inversely with the permittivity and permeability of the media:

$$D_{1n} = D_{2n}; \ \frac{E_{1n}}{E_{2n}} = \frac{\varepsilon_2}{\varepsilon_1}, \tag{1.16}$$

$$B_{1n} = B_{2n}; \ \frac{H_{1n}}{H_{2n}} = \frac{\mu_2}{\mu_1} \tag{1.17}$$

where ε_1 and μ_1 are the parameters of the first medium and ε_2 and μ_2 are the parameters of the second medium.

The tangent components of the field vectors are constant, while the tangent components of the field induction vectors change proportionally to the permeability.

$$\frac{D_{1\pi}}{D_{2\pi}} = \frac{\varepsilon_1}{\varepsilon_2}; E_{1\tau} = E_{2\tau} \tag{1.18}$$

$$\frac{B_{1\tau}}{B_{2\tau}} = \frac{\mu_1}{\mu_2}; H_{1\tau} = H_{2\tau} \tag{1.19}$$

Surface charges ρ' and surface currents j' arise on the surface of an electric conductor. Surface charges change only the normal component of the electric induction vector $D_{1n} - D_{2n} = \rho'$, and surface currents change only the perpendicular to the current tangential component of the magnetic field $H_{1\tau} - H_{2\tau} = j_N'$.

In an ideal conductor, electric conductivity $\sigma \to \infty$ and alternating electromagnetic fields cannot exist – an ideal conductor short-circuits the alternating electric field. Also, there cannot be an electrostatic field inside an ideal conductor – the charges are distributed on the surface of the conductor in such a way that the total field in the conductor is equal to zero. Otherwise, these charges would move forward under the influence of the field. Thus, in the case of an ideal conductor, all field components become zero in the second environment $E_{2n} - E_{2\tau} = H_{2n} = H_{2\tau} = 0$. Following boundary conditions, the components of the electromagnetic field in the first medium are related to the surface charges and surface currents. The boundary condition on the surface of an ideal conductor:

$$E_{1n} = \frac{\rho'}{\varepsilon_0 \varepsilon_1} \tag{1.20}$$

$$E_{1\tau} = 0 \tag{1.21}$$

$$H_{1n} = 0 \tag{1.22}$$

$$H_{1\tau} = j_N \tag{1.23}$$

An important conclusion from the boundary conditions is that on the surface of an ideal as well as nonideal electric conductor, the electric field is perpendicular to the boundary surface and the magnetic field is parallel to the surface (Figure 1.7). The tangential component of the magnetic field is equal to the surface current density transverse to it (Eq. 1.23).

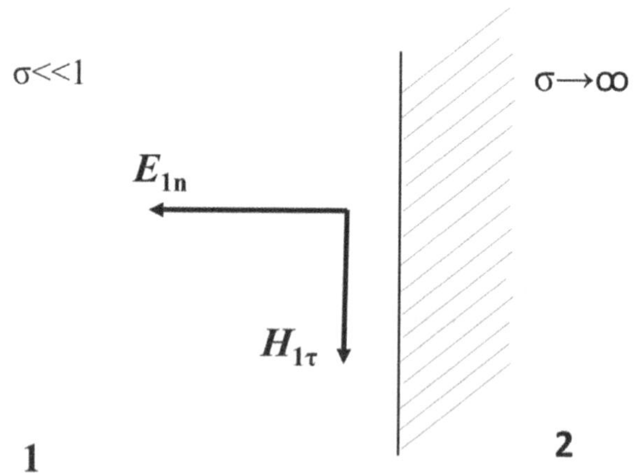

FIGURE 1.7 The electromagnetic field structure in a dielectric 1 at the boundary of an ideal conductor 2.

To fulfill the boundary condition for a conductive surface parallel to the electric field, the horizontal component of the reflected electric vector must be equal and opposite to that of the incident wave. These conditions affect field structure and should be considered also on the border of nonideal conductors, as a medium with a dielectric constant much higher than one.

It is important to consider the boundary conditions in planning experiments to evaluate the effects of electromagnetic fields on biological objects. For example, the radiation from the vertical horn antenna that produces an electric field parallel to its aperture is "rejected" from the samples on horizontal Petri dishes. The vertical solenoid produces the vertical magnetic field "rejected" by horizontal samples. Neglecting the boundary conditions can cause misleading results in the performed experiments.

Boundary conditions, derived from Maxwell's equations, are universal and applicable at all frequencies in all kinds of media and systems and under all kinds of conditions. It is not possible to change them. Thus, the structure of the field in the electromagnetic wave in any system is such that the boundary conditions are automatically fulfilled. In a metal system, the electric field should be directed perpendicular to the metal. If there is also a component of the electric field parallel to the metal, it becomes zero near the metallic surface. In this way, a wave with a more complex structure can be created in the system where minimums and maximums occur.

1.7 ENERGY OF ELECTROMAGNETIC FIELD

Poynting's theorem, fixing the energetic relations in the electromagnetic field, states the law of conservation of energy in the electromagnetic field. The law of conservation of energy is derived from the first and second Maxwell equations by multiplying them scalarly by $\boldsymbol{E}$ and $\boldsymbol{H}$, respectively, and subtracting. The further application of the vector analysis transformation $\nabla(\boldsymbol{A}\times\boldsymbol{B})=\boldsymbol{B}(\nabla\times\boldsymbol{A})-\boldsymbol{A}(\nabla\times\boldsymbol{B})$ results in the equation

$$\Delta(\boldsymbol{E}\times\boldsymbol{H})+\left(\boldsymbol{E}\frac{d\boldsymbol{D}}{dt}+\boldsymbol{H}\frac{d\boldsymbol{B}}{dt}\right)+\boldsymbol{Ej}=0 \tag{1.24}$$

The Poynting's vector

$$\boldsymbol{P}=\boldsymbol{E}\times\boldsymbol{H} \tag{1.25}$$

presents the power density of an electromagnetic field and describes the flow of electromagnetic energy.

Alternating electric and magnetic fields are harmonic functions of frequency. Their product differs from zero if both field vectors differ from zero. In the case of a 90-degree phase shift between the oscillating vectors, one of these is equal to zero once in each quarter-period. This is the case of a reactive field, the electric energy turns into magnetic energy, and after a quarter-period the magnetic energy turns into electric energy (Figure 1.8). The energy is oscillating from electric to magnetic and vice versa.

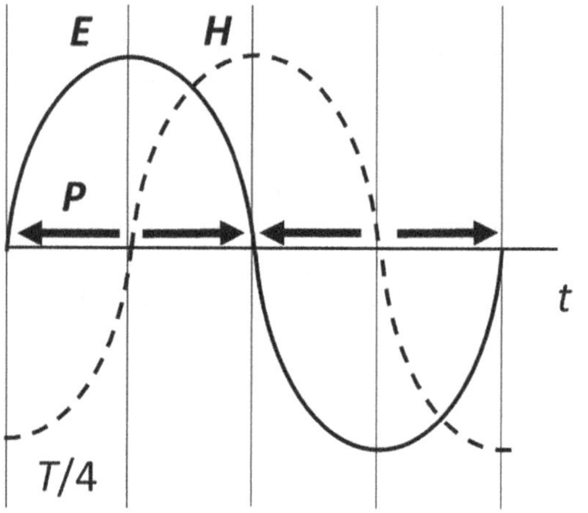

FIGURE 1.8 Field distributions in a standing wave at a fixed moment t = const.

The energy of a standing wave is reactive. The energy oscillates within a half-wave and does not propagate in space. The reactive energy is in electromagnetic oscillators, like *LC* circuits and radiofrequency resonators, used as frequency-selective elements in amplifiers, generators, and filters.

Considering equation (1.24) in the space V enclosed by the surface S and equations (1.5, 1.7, and 1.9) leads to the result

$$\int \boldsymbol{P}d\boldsymbol{S} + \frac{d}{dt}\int \varepsilon \boldsymbol{E}^2 + \mu \boldsymbol{H}^2)dV + \int \sigma \boldsymbol{E}^2 dV = 0 \tag{1.26}$$

In equation (1.26)

$$w_E = \varepsilon \boldsymbol{E}^2 \tag{1.27}$$

is the electric energy density. The magnetic energy density

$$w_H = \mu \boldsymbol{H}^2 \tag{1.28}$$

The expression

$$p_T = \sigma \boldsymbol{E}^2 \tag{1.29}$$

presents the power density of losses. This equation is the base of calculation specific absorption rate (SAR), the parameter used for evaluating the exposure level considering thermal effects

$$SAR = \frac{\sigma \boldsymbol{E}^2}{m_d} \tag{1.30}$$

where m_d is the mass density.

The first term in equation (1.26) describes the flow of electromagnetic energy through the surface S that can be positive or negative, increasing or decreasing the energy within the surface. The electromagnetic energy flow through the surface S is described only by the real part of a Poynting vector. The imaginary part of the vector describes reactive oscillating energy, changing the direction of its movement twice during a period with zero average flow. The second term describes the alteration of total electromagnetic energy, the sum of electric and magnetic energies, in space V within the surface S. The third term corresponds to the thermal losses in space V according to Joule's law I^2R. The law of conservation of electromagnetic energy (Eq. 1.26) states the balance between the alterations of total electromagnetic energy in space without additional sources of energy, the losses of energy within the space, and the flow of energy through the surface surrounding the space.

1.8 CONCLUSIONS

The conclusions are formulated considering their relevance to understanding the mechanism of electromagnetic field effects in the brain.

- The electric field can affect stationary and moving charged particles; the magnetic field affects only moving charged particles.
- Alternating electric and magnetic fields are connected and perpendicular. One cannot exist without the other.
- The impact of a medium on the properties of an electromagnetic field is described by dielectric polarization.
- Dielectric polarization caused by the impact of an external electric field on charged particles in a medium depends linearly on the electric field strength. The effect of electric field to a medium is determined by the electric field strength, not SAR.
- The Maxwell field equations are linear. In a linear medium, different electromagnetic fields behave independently and do not affect each other.
- The electric field is perpendicular to a conductor; the magnetic field is parallel to the surface of the conductor.

REFERENCES

King RWP, Smith GS, (1981). *Antennas in matter: fundamentals, theory and application.* MIT Press, Cambridge, MA.

Maxwell JC, (1873). *A treatise on electricity and magnetism.* Oxford University Press, London.

Ramo S, Whinnery R, (1944). *Fields and waves in modern radio.* Wiley, New York.

Ramo S, Whinnery JR, Van Duzer T, (1965 and 1994). *Fields and waves in communication electronics.* Wiley, New York.

Zahn M, (1979). *Electromagnetic field theory, a problem solving approach.* John Wiley & Sons, New York.

2 Electromagnetic radiation and harmonic coherent waves in an inhomogeneous medium

HIGHLIGHTS

- Due to the linearity of the wave equation, the sum of two or more coherent waves in a linear medium is coherent.
- Wavelength and phase velocity depend on the dielectric permittivity of the medium.
- The electric field is directed perpendicular to the conductor, and the magnetic field is parallel to it.
- The direction of wave propagation and polarization of the electric field can be changed upon reflection, but not its coherence.
- Poynting vector, the vector product of electric and magnetic fields, describes the propagation of a wave and the transfer of radiation energy. The Poynting vector is imaginary, and there is no transfer of wave energy in a standing wave (the phase shift between the electric and magnetic vectors is 90°).
- No antenna radiates directly down.

2.1 INTRODUCTION

The theory of natural thermal radiation was developed by Max Planck (1900). He introduced the concept of the discrete nature of electromagnetic radiation. Planck assumed that energy could not be absorbed continuously but only in discrete, small amounts known as quanta. The Nobel Prize in Physics 1918 was awarded to Max Planck for his discovery of quantum energy. Einstein introduced the more general concept of wave-particle duality to physics in 1909. Quantum is the smallest amount of energy carried by photons, radiation particles with zero mass. The spectral density of radiation energy w_f is proportional to radiation frequency f as $w_f = hf$, where h is Planck's constant, $h = 6.626\times10^{-34}$ joule-second.

Thermal radiation depends not only on frequency but also on the temperature of the radiating body. The energy density of black body radiation w_T is proportional to the absolute temperature of the radiating body T as $w_T = kT$, where $k = 1.380649\text{e}10^{-23}$ joule per kelvin is the Boltzmann constant. Each body with a temperature higher than absolute zero (0 K) irradiates thermal radiation. At a temperature of 300 K, the spectral energy of radiation quanta becomes equal to

DOI: 10.1201/9781003729976-3

thermal energy at the frequencies of about 10^{13} Hz in the infrared part of the spectrum at micrometers wavelengths.

Planck's law determines the nonmonotonic relationship between the radiation energy density and frequency, which varies at different temperatures. The spectral density of thermal radiation $w(f, T)$ varies with frequency f and the temperature of the radiating body T as

$$w(f,T)=\frac{2f^2}{c^2}\frac{hf}{e^{\frac{hf}{kT}}-1} \tag{2.1}$$

The coefficient $2f^2/c^2$ in the formula (2.1) determines the number of oscillation modes. Despite a strong tendency to increase with frequency f^3, the dependence of power density on frequency is nonmonotonic, indicating a decrease at higher frequencies due to $\exp(hf/kT)$ in the denominator. The maximum of the thermal radiation spectrum occurs at the condition $hf = kT$ and shifts to higher frequencies as the temperature increases (Figure 2.1). The maximum of the Sun's thermal radiation spectrum ($T = 5770$ K) is in visible light wavelengths. The maximum of the human body thermal radiation spectrum ($T = 300$ K) is much lower and occurs at infrared frequencies.

For single-mode radiation at radiofrequencies, the formula (2.1) can be simplified considering $hf \ll kT$. The spectral energy density of thermal radiation at radiofrequencies is proportional to temperature $w(f, T) = kT$.

Each body at a temperature higher than 0 K emits thermal radiation. The spectral power density of thermal radiation from non-black body $P(f, T)$ is proportional to the temperature of the body and its ability to radiate, $P(f, T) = \alpha T$, where α is the radiation coefficient of the body equal to its absorption coefficient, and T is the temperature of the body. The equivalent radiation temperature $T_r = \alpha T$ determines the radiation from a non-black body. Whereas $\alpha < 1$, the maximum of radiation occurs at lower frequencies compared to radiation from an ideal blackbody.

The history of technical electromagnetic radiation grows from the development of electromagnetism. The first principal idea about the propagation of electromagnetic disturbances as a wave was raised by James Clerk Maxwell (1846). Later, he

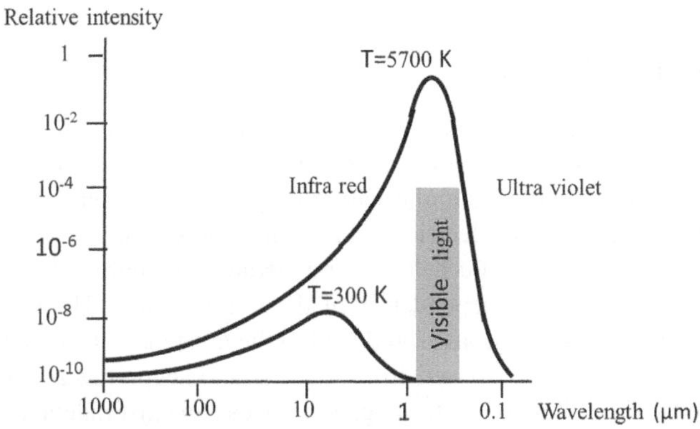

FIGURE 2.1 The distribution of the power spectral density of thermal radiation.

generalized this idea and formulated the mathematical fundamentals of electromagnetism, which were published in the book *A Treatise on Electricity and Magnetism* in 1873. The four Maxwell equations describe the fundamental laws of electromagnetism without any limitations on the frequency range. Mathematically, he derived the general equations that explained all the various phenomena of electricity and magnetism, binding them together. The first two Maxwell equations predicted the existence of electromagnetic radiation. He has theoretically demonstrated that the electromagnetic field produced by an oscillating electric charge is moving outward from its source at a constant speed. Heinrich Rudolf Herz confirmed Maxwell's predictions in his experiments conducted between 1886 and 1888. Herz used an oscillating electric spark to induce electromagnetic oscillations in a distant wire loop. Interestingly, the first application of high-frequency electromagnetic fields was in the field of medicine and biology. Jaques-Arsene d'Arsonval studied a wide variety of physiological effects of alternating currents and time-varying electric and magnetic fields at high frequency in 1891–1893. The diathermy method introduced by d'Arsonval is used in electrotherapy today. The first developments in telecommunication engineering were carried out by Guglielmo Marconi (Marconi, 1899). Somewhat later, he began radio experiments in 1894 and demonstrated the transfer of a radio signal over a distance of 19 km in 1897. The Nobel Prize in Physics 1909 was awarded to Guglielmo Marconi for contributions to the development of wireless telegraphy.

The theory of electromagnetic waves generated by harmonic oscillators-generators at radio frequencies regarding telecommunication technology, presented in the classical book by Ramo, Whinnery and van Duzer (Ramo et al., 1994) and in the modern textbook by Zahn (Zahn, 1979) is based on Maxwell's equations and mathematically derived from these.

This chapter outlines the aspects of electromagnetic radiation that are important for understanding the radiation effects on humans. The properties of harmonic electromagnetic waves enable us to describe wave propagation in homogeneous and inhomogeneous media and the spatial distribution of field power density. This knowledge significantly contributes to the correct determination of exposure level and design of exposure equipment.

2.2 WAVE EQUATION

An electromagnetic wave is an electromagnetic disturbance that propagates in space with finite velocity. The possibility of electromagnetic radiation arises from the first two Maxwell's equations: an alternating magnetic field creates an electric field (Eq. 1.12), and an alternating electric field in turn creates a magnetic field (Eq. 1.13). The alternating electromagnetic field, once generated by charges (Eq. 1.14) or currents (Eq. 1.12), can exist independently of the initial source. Heinrich Hertz first realized that Maxwell's equations would lead to the existence of radio waves.

A wave equation can be derived from the first and second Maxwell's equations (Eqs. 1.12 and 1.13). Whereas electromagnetic waves can exist independently of their source, we exclude sources $\boldsymbol{J}$ and ρ from the system. Inconveniently, both equations include both electric and magnetic fields. It is possible to express the vector $\boldsymbol{H}$ in the first equation in terms of the vector $\boldsymbol{E}$ from the second equation and vice versa. Using the identities of vector analysis, the wave equations can be derived separately for the electric and magnetic components of the field:

$$\nabla^2 \boldsymbol{E} - \varepsilon\mu \frac{\partial^2 \boldsymbol{E}}{\partial t^2} = 0 \tag{2.2}$$

$$\nabla^2 \boldsymbol{H} - \varepsilon\mu \frac{\partial^2 \boldsymbol{H}}{\partial t^2} = 0. \tag{2.3}$$

Two wave equations and their general solutions are identical for electric and magnetic fields. Therefore, when discussing the general properties of electromagnetic waves, only the equation for the electric field is considered. The differences in the boundary conditions between electric and magnetic fields can cause differences between the detailed solutions for $\boldsymbol{E}$ and $\boldsymbol{H}$. Considering the specific systems, $\boldsymbol{E}$ and $\boldsymbol{H}$ can have different distributions.

Technical sources of radiofrequency radiation generate harmonic fields at a fixed frequency f. Pulsed-modulated fields used in telecommunication can be considered as a combination of harmonic waves. The limited range of modulation frequencies compared to the carrier radiofrequency makes it possible to consider the radiation used in telecommunication as harmonic. Therefore, the theory of harmonic waves is also suitable for describing modulated radiofrequency radiation. In the case of harmonic waves, the time dependence becomes exp $(i\omega t)$, where $\omega = 2\pi f$ is the angular frequency, and so $\partial^2 / \partial t^2 = -\omega^2$. A source generates a spherical wave that propagates with phase velocity v from the source. The phase surface of the wave is a sphere with a radius of r, indicating distance from the source. The wave depends only on one coordinate, distance from the source r. Far from the source, the curve of the sphere is minimal, and the wave can be considered a plane wave. The phase plane of a plane wave is perpendicular to the propagation direction, the fields on the phase plane are constant, and $\frac{\partial}{\partial x} = \frac{\partial}{\partial y} = 0$. For harmonic plane waves, the wave equations (Eqs. 2.1 and 2.2) can be written in a more convenient form:

$$\frac{\partial^2 \boldsymbol{E}}{\partial r^2} + k^2 \boldsymbol{E} = 0, \tag{2.4}$$

where $k = \omega\sqrt{\varepsilon\mu} = \frac{\omega}{v} = \frac{2\pi}{\lambda}$ is the wave number or propagation (phase constant), v is the phase velocity (the velocity of transfer of the phase plane), and λ is the wavelength. The phase velocity is determined by the parameters of the medium

$$v = \frac{1}{\sqrt{\varepsilon\mu}} = \frac{c}{\sqrt{\varepsilon_r \mu_r}} \tag{2.5}$$

where $c = 1/\varepsilon_0\mu_0$ is the phase velocity in vacuum.

The solution of the wave equation (Eq. 2.3) is:

$$\boldsymbol{E} = \boldsymbol{E}_o \cos(\omega t - kr) = \boldsymbol{E}_o \cos\left(t - \frac{r}{v}\right) \tag{2.6}$$

where $\boldsymbol{E}_0$ is the amplitude of the electrical field.

The wave propagates with a determined velocity v in space along r. The second term in the cosine argument determines the phase distribution in space. It is the phase shift relative to the initial phase of the source at a distance r from the source. The initial phase at the source of radiation is assumed to be constant and is not indicated in formulas. The sign in the cosine's argument presents a phase shift in the propagation direction and defines the direction of wave propagation. The negative sign shows the delay of the phase with increasing distance r and defines the propagation from $r = 0$ to $r \to \infty$. Positive sign defines phase advance and the propagation in an opposite direction, from $r \to \infty$ to $r = 0$.

The fact that the phase velocity is a definite quantity and not infinitely large is of fundamental importance for the existence of electromagnetic radiation. This fact transfers the temporal dependence in the time domain to the dependence on the coordinate in the spatial domain. In the case of the infinitely high phase velocity, the fixed constant phase is distributed from $r = 0$ to $r \to \infty$, and the wave becomes a quasi-static field.

2.3 PLANE ELECTROMAGNETIC WAVE IN AN IDEAL ISOTROPIC HOMOGENEOUS MEDIUM

In an isotropic medium, the parameters of the medium ε and μ are scalars and the properties of the medium don't depend on the direction of wave propagation. An ideal medium is considered a medium without electric or magnetic losses. Such a medium is a vacuum. Very close to the ideal homogeneous medium is the atmosphere. The relative electric permittivity of the atmosphere near the Earth's surface is 1.0006, and the magnetic permeability is 1. This is valid for a clear atmosphere. Rain, fog, snow, dust, and pollution can cause remarkable losses of electromagnetic energy in the atmosphere.

In an ideal medium, the amplitude of an electromagnetic wave is constant. Electric and magnetic vectors, being perpendicular to each other, are oscillating in different planes. The vector product of the electric and magnetic field vectors determines the direction of wave propagation (Figure 2.2). The Poynting vector $\boldsymbol{P}$

$$\boldsymbol{P} = \boldsymbol{E} \times \boldsymbol{H} \tag{2.7}$$

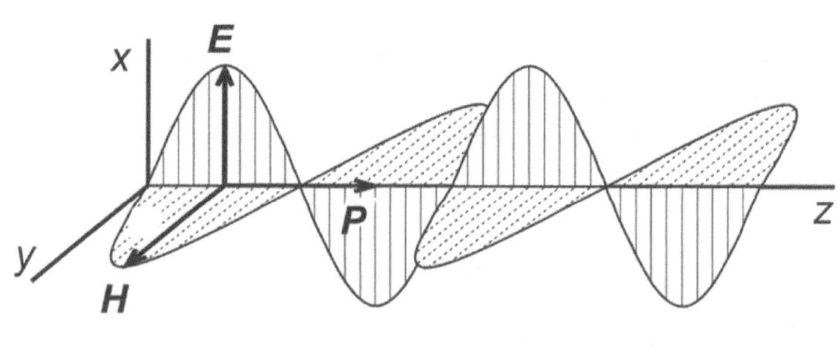

FIGURE 2.2 A plane electromagnetic wave in an ideal homogeneous isotropic medium.

represents the power flow of electromagnetic energy.

In a lossless medium, the electrical permittivity ε and magnetic permeability μ are real scalar values, and the wavenumber k is also real. The phase velocity is a simple function of the parameters of the medium. The wavelength is the distance traveled during a period of $T = 1/f$. The phase velocity determines the wavelength λ, which depends on the parameters of the medium as

$$\lambda = vT = \frac{v}{f} = \frac{1}{f\sqrt{\varepsilon\mu}} = \frac{c}{f\sqrt{\varepsilon_r\mu_r}} = \frac{\lambda_0}{\sqrt{\varepsilon_r\mu_r}} \tag{2.8}$$

where λ_0 is the wavelength in a vacuum.

The main properties of the plane wave in an ideal medium can be defined by considering the wave equation and Maxwell's equations.

1. The amplitude of the plane wave in a lossless medium is constant, and the wave is uniform.
2. The electrical and magnetic field vectors oscillate in perpendicular planes and are perpendicular to each other $\boldsymbol{E} \perp \boldsymbol{H}$.
3. The ratio of the total electric field to the total magnetic field is the wave impedance of the free space:

$$W = \frac{E}{H} = \sqrt{\frac{\mu}{\varepsilon}} \tag{2.9}$$

4. The Poynting vector $\boldsymbol{P}$ determines the transfer of the energy of the electromagnetic wave:

$$\boldsymbol{P} = \boldsymbol{E} \times \boldsymbol{H} = \frac{\boldsymbol{E}^2}{W}\boldsymbol{n} = \boldsymbol{H}^2 W\ \boldsymbol{n} \tag{2.10}$$

   A unit vector $\boldsymbol{n}$ is directed in the direction of wave propagation. Only the components of the vectors $\boldsymbol{E}$ and $\boldsymbol{H}$, perpendicular to the wave propagation direction, contribute to the transmission of the wave power density.
5. The direction of the wave propagation is determined by the direction of the Poynting vector. Its direction changes if one of the vectors $\boldsymbol{E}$ or $\boldsymbol{H}$ changes its phase.

It is important to keep in mind that the plane wave is an approximation, and a real wave in the atmosphere is spherical. The electromagnetic energy of a spherical wave is distributed over the sphere of radius r, and the average power density p depends on the distance from the source.

$$p = \frac{P_0}{4\pi r^2} \tag{2.11}$$

where P_0 is the power generated by the source. Despite the amplitude of an ideal plane wave being constant in an ideal medium, the amplitude of a real wave decreases with distance. Whereas the power density $p \sim \frac{1}{r^2}$, the amplitude of the wave

$$E_0 \sim \frac{1}{r} \tag{2.12}$$

2.4 PLANE ELECTROMAGNETIC WAVE IN A REAL ISOTROPIC HOMOGENEOUS MEDIUM

A nonideal real medium is a lossy medium. In most materials, electric phenomena cause the losses. Electric losses are related to conduction currents. Combining the displacement current i$\omega\boldsymbol{E}$ and the conduction current $\sigma\boldsymbol{E}$, the electric permittivity becomes complex:

$$\varepsilon_c = \varepsilon\left(1 - i\frac{\sigma}{\omega\varepsilon}\right) = \varepsilon' - i\varepsilon'' \tag{2.13}$$

The first component in the formula describes the effect of dielectric polarization, and the second component describes the impact of ohmic losses. At higher frequencies $\omega\varepsilon >> \sigma$, the impact of conductivity on the absolute value of dielectric polarization is negligible.

In a lossy medium, the electric permittivity depends on the frequency. The losses in a medium are related to the conduction current. The relative impact of losses is described by the ratio of conduction current to displacement current, named loss tangent:

$$\tan\delta = \frac{\sigma}{\omega\varepsilon} \tag{2.14}$$

In a real (non-perfect) dielectric medium, the conductive current is much less than the displacement current, $\sigma << \omega\varepsilon$ and tan $\delta << 1$. The displacement current dominates at high frequencies at $\omega >> \sigma/\varepsilon$. In a good conductor, the conductive current is much larger than the displacement current, $\sigma >> \omega\varepsilon$ and tan $\delta >> 1$. The conduction current dominates at lower frequencies. The same medium can be a good conductor at lower frequencies and a good dielectric at higher frequencies. For example, seawater is a good conductor at low frequencies but behaves as a dielectric at microwaves and is transparent to visible light.

The complex electric permittivity also causes complexity of the wavenumber k, represented by a complex propagation constant for the medium:

$$\gamma = \beta - i\alpha = k\sqrt{1 - i\frac{\sigma}{\omega\varepsilon}} \tag{2.15}$$

The phase constant β and attenuation constant, as a real and imaginary part of the complex propagation constant, is calculated from the equation above:

$$\beta = k\sqrt{\frac{1}{2}\left(\sqrt{1 - \left(\frac{\sigma}{\omega\varepsilon}\right)^2} + 1\right)} \tag{2.16}$$

$$\alpha = k\sqrt{\frac{1}{2}\left(\sqrt{1-\left(\frac{\sigma}{\omega\varepsilon}\right)^2}-1\right)} \tag{2.17}$$

There are principal differences in wave propagation in an ideal and nonideal medium. First, the plane wave is no longer uniform, and the attenuation constant describes the rate of decay of its amplitude. An exponential damping factor appears in the solution of the wave equation:

$$\boldsymbol{E} = \boldsymbol{E}_0 e^{\alpha z}\cos(\omega t - \beta z) \tag{2.18}$$

The second principal difference between the wave propagation in ideal and nonideal media is the dependence of phase velocity on frequency in a nonideal medium. The frequency dependence of the phase velocity is induced by the phase constant

$$v_f = \frac{\omega}{\beta} = \frac{v_0}{\sqrt{\frac{1}{2}\left(\sqrt{1-\left(\frac{\sigma}{\omega\varepsilon}\right)^2}+1\right)}} \tag{2.19}$$

where v_0 is a propagation velocity in an ideal medium, where $\sigma = 0$.

The dependence of the propagation constant on the frequency causes dispersion. A lossy medium is dispersive: every spectral component of the polychromatic wave has its own phase and group propagation velocity. The phase velocity describes the propagation of the phase front. The group velocity v_g is the velocity of energy propagation. The phase and group velocities are related to each other with a convenient relationship for practical use:

$$v_f v_g = v_0^2 \tag{2.20}$$

The group velocity is limited to the value of v_0, but the phase velocity can exceed v_0.

In a real conductor $\sigma >> \omega\varepsilon$ and the phase and attenuation constants can be expressed by an approximate relation from the formulas (2.14) and (2.15), respectively:

$$\alpha = \beta = \sqrt{\frac{\omega\mu\sigma}{2}} \tag{2.21}$$

The magnitude of the field decreases exponentially as it penetrates the conductive medium. The value of the depth at which the field has decreased to $1/e$ of its value at the surface is called the depth of penetration (depth of the skin layer) and is determined as

$$\Delta = \frac{1}{\alpha} = \sqrt{\frac{2}{\omega\mu\sigma}} \tag{2.22}$$

As the magnitude of the field decreases exponentially, the wave propagates deeper than the depth of penetration.

An electromagnetic wave cannot penetrate an ideal conductor: when $\sigma \to \infty$, the depth of penetration $\Delta \to 0$.

It is important to point out that the depth of penetration depends not only on the parameters of the medium but also on the frequency of the electromagnetic wave. The depth of penetration decreases with frequency.

2.5 PLANE ELECTROMAGNETIC WAVE IN AN ISOTROPIC INHOMOGENEOUS MEDIUM

In an isotropic inhomogeneous medium, the parameters of the medium ε and μ are scalars and have a nonuniform spatial distribution. Two cases of the distribution of parameters inside the medium can be discussed:

1. A step change of the parameters on the border between two media with different parameters
2. A continuous change of the parameters inside a medium

In the case of a step change of the parameters, the reflection and refraction of the plane wave take place at the border between the media (Figure 2.3). According to Snell's law of refraction, the change of the propagation direction on the border between two dielectric (nonmagnetic) media is

$$\sqrt{\varepsilon_1}\sin\theta_1 = \sqrt{\varepsilon_2}\sin\theta_2 \tag{2.23}$$

where θ_1 is an incident angle, θ_2 is a refraction angle, ε_1 is the permittivity of the first medium, and ε_2 is the permittivity of the second medium.

In the case of propagation from a dielectric medium to the real conductor $\varepsilon_1 << |\varepsilon_2|$ and, independently of the incident angle θ_1, the refraction angle $\theta_2 \to 0$. Inside a real conductor, a plane wave propagates perpendicularly to the surface of a medium.

2.5.1 Reflection on the Border between Two Media

The reflection coefficient R is the ratio of the reflected wave amplitude to the incident wave amplitude. The coefficient R can be derived from boundary conditions on the border between two media. Whereas the boundary conditions are different for the normal and tangent components of field vectors, coefficient R differs for the different orientations of the electric vector $\boldsymbol{E}$ relative to the border. The plane of oscillations of the vector $\boldsymbol{E}$ determines the polarization of the wave. The reflection coefficients in the case of vertical polarization R_v and horizontal polarization R_h of the wave are consequently:

$$R_v = \frac{\varepsilon_2\cos\theta_1 - \sqrt{\varepsilon_1}\sqrt{\varepsilon_2 - \varepsilon_1\sin^2\theta_1}}{\varepsilon_2\cos\theta_1 + \sqrt{\varepsilon_1}\sqrt{\varepsilon_2 - \varepsilon_1\sin^2\theta_1}} \tag{2.24}$$

$$R_h = \frac{\sqrt{\varepsilon_1\cos\theta_1} - \sqrt{\varepsilon_1}\sqrt{\varepsilon_2 - \varepsilon_1\sin^2\theta_1}}{\sqrt{\varepsilon_1\cos\theta_1} + \sqrt{\varepsilon_1}\sqrt{\varepsilon_2 - \varepsilon_1\sin^2\theta_1}} \tag{2.25}$$

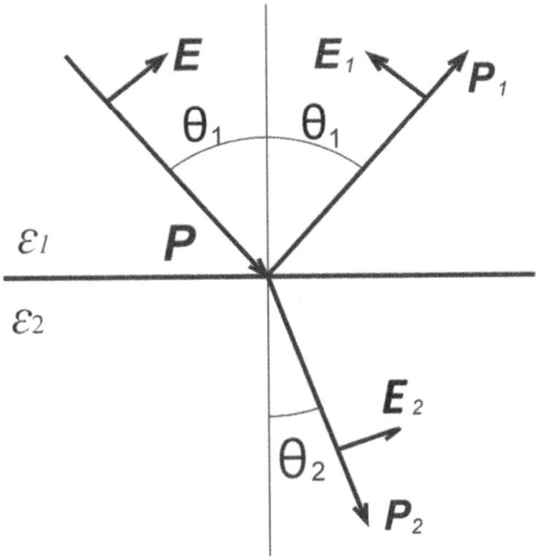

FIGURE 2.3 Reflection and refraction of a plane wave at the border of two media.

The electric permittivity is a complex quantity (Eq. 2.3). Therefore, the reflection coefficients are complex. Not only the amplitude but also the phase of the reflected wave differs from the phase of the incident wave.

In the case of wave transmission from the medium with lower permeability to a medium with higher permeability $\varepsilon_1 < \varepsilon_2$ (from air to tissue) the reflection coefficient for vertical polarization becomes equal to zero when

$$\varepsilon_2 \sin\theta_{1B} = \sqrt{\varepsilon_1}\sqrt{\varepsilon_2 - \varepsilon_1 \sin^2\theta_{1B}} \tag{2.26}$$

The angle θ_{1B} is the angle of full refraction (Brewster angle), and all energy of the electromagnetic wave is transferred into the second medium without reflection. This angle is important for the optimization of the electromagnetic wave energy transmission between different dielectric media. When the second medium is a good conductor, the Brewster angle becomes close to 90° and it is difficult to achieve the effect of full refraction. The Brewster angle does not exist for horizontal polarization.

The value of the reflection coefficient increases with an increase in the incident angle for vertical and horizontal polarization. In the case when $\theta_1 \to 90^\circ$, the module of the reflection coefficient $|R| \varnothing 1$.

The value of the reflection coefficient decreases with a decrease in the incident angle for both polarizations. Therefore, the optimal incident angle for transferring electromagnetic wave energy to a medium is equal to zero. The reflection coefficient of a plane wave incident from free space (air, dielectric) perpendicularly onto the border of a conductive nonmagnetic medium results as

$$R = \frac{\sqrt{\varepsilon_2} - \sqrt{\varepsilon_1}}{\sqrt{\varepsilon_2} + \sqrt{\varepsilon_1}} = \frac{W_1 - W_2}{W_1 + W_2} \tag{2.27}$$

where W_1 and W_2 are the wave impedances of the first and second media.

2.5.2 Standing Wave

The boundary condition on the surface of an ideal conductor (Eqs. 1.20–1.23) determines the field structure of electric and magnetic fields near an ideal conductor (Figure 2.4). The reflected wave interferes with the incident wave. The full reflection back when $|R| = 1$ results in a standing wave as a sum of forward and back-propagating waves $\boldsymbol{E}_1$ and $\boldsymbol{E}_2$ with equal amplitudes

$$\boldsymbol{E} = \boldsymbol{E}_1 + \boldsymbol{E}_2 = \boldsymbol{E}_0\left[(\cos\omega t - kr) + (\cos\omega t + kr)\right] = 2\boldsymbol{E}_0 \cos kr \cos\omega t \quad (2.28)$$

where $\boldsymbol{E}_0$ is the amplitude of the incident wave. Standing waves do not transfer electromagnetic wave energy. The character of the electromagnetic field becomes reactive: the electric and magnetic field vectors have a 90° phase shift to each other, and the Poynting vector becomes imaginary (Section 1.6). This is an oscillating field: electric field energy is transferred to magnetic and vice versa. The fields achieve their maximum at a quarter-period time interval, and these are located at a quarter wavelength distance from each other (Figure 2.4).

In the case of reflection from a nonideal conductor $|R| < 1$ the amplitude of the reflected wave in equation (2.27) is smaller than the amplitude of the incident wave, $\boldsymbol{E}_{20} = \boldsymbol{E}_{10}(1 - R)$. A partial reflection back results in a partial standing wave. The partial standing wave in a medium can be described as

$$\boldsymbol{E} = \boldsymbol{E}_1 + \boldsymbol{E}_2 = \boldsymbol{E}_0(1 - R)(\cos\omega t - kr) + \boldsymbol{E}_0 R(\cos\omega t + kr) \quad (2.29)$$

The electric field of a partial standing wave varies between $\boldsymbol{E}_0(1 - R)$ and $\boldsymbol{E}_0(1 + R)$. The part of the wave with constant amplitude $\boldsymbol{E}_0(1 - \mathrm{R})$ transfers active electromagnetic energy. The wave with amplitude $\boldsymbol{E}_0 R$ describes oscillating reactive energy.

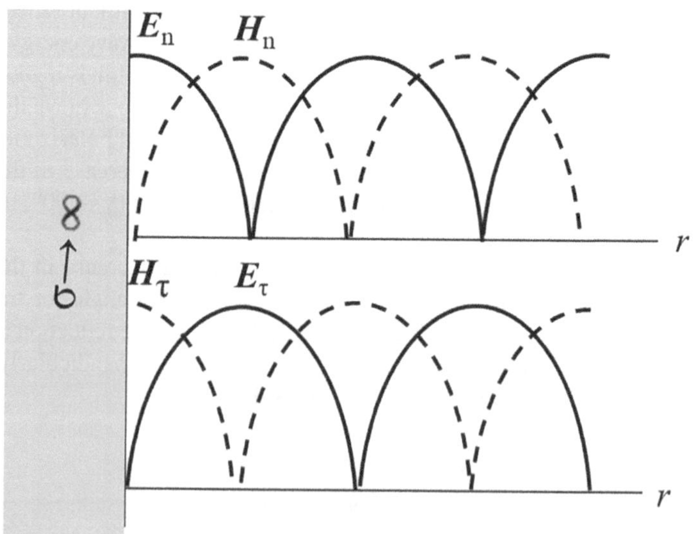

FIGURE 2.4 The structure of electric and magnetic fields near an ideal conductor.

Metallic surfaces can be considered as ideal conductors. The human body and biological tissues behave more as nonlinear conductors than dielectrics due to relatively large values of dielectric constants (Section 3.1).

2.5.3 Refraction in an Inhomogeneous Medium

The reflection at the interface between two media decreases with a decrease in the differences between the parameters of these media. As a limit of such decrease in the differences, the continuous change of the permittivity, instead of the boundaries between the media, can be considered. The reflection becomes negligible when the relative change of permeability over a distance of a wavelength inside the medium is small. This is the case of full refraction of a wave propagating in an inhomogeneous medium.

A continuous change of the parameters inside an inhomogeneous medium can be considered using a multilayered model. The continuous change of parameters is replaced by a combination of the elementary layers (Figure 2.5). The parameters of the medium have a constant value inside each layer. The thicknesses of the layers are selected to be smaller than the wavelength to fulfill the criteria of full refraction. The equation for the wave propagation trajectory inside the multilayered medium is derived from the refraction law as

$$\sqrt{\varepsilon_1}\sin\theta_1 = const \tag{2.30}$$

The increase in value of the permittivity ε_i causes a decrease in the angle θ_i and vice versa. Therefore, the trajectory of the wave propagation declines toward the higher values of the dielectric permeability.

2.6 POLARIZATION OF ELECTROMAGNETIC WAVES

2.6.1 Linear, Circular, and Elliptic Polarization

Discussing an electromagnetic wave, traditionally we assumed the linear polarization of the wave where the electric vector oscillates in a fixed plane and the magnetic field

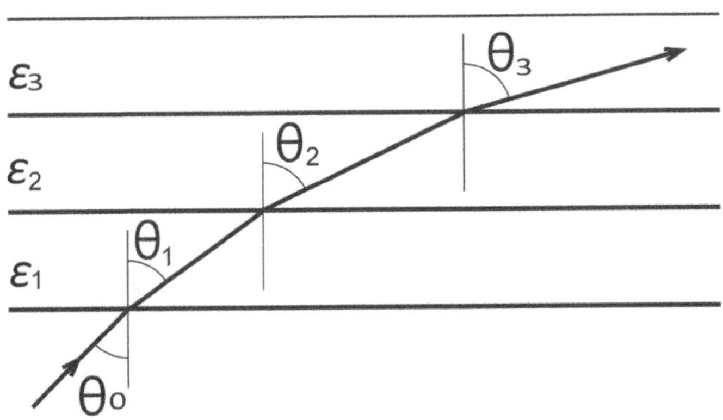

FIGURE 2.5 Refraction of an electromagnetic wave in a medium with continuous change of the parameters.

is perpendicular to that (Figure 2.2). The orientation of the electric field determines the plane of polarization. The plane of polarization in space depends on the orientation of currents and fields on the transmitting antenna. However, the polarization of an electromagnetic wave is not always linear because the electric field can oscillate in different planes. The polarization can be not only linear but also circular or elliptic.

In general, the electric field of a wave can oscillate in an arbitrary plane. A linearly polarized electromagnetic wave can be presented as a sum of two partial waves polarized in two perpendicular planes XZ and YZ, and the electric field of a linearly polarized wave $\boldsymbol{E}$ can be divided into two perpendicular components, $\boldsymbol{E}_x$ and $\boldsymbol{E}_y$

$$\boldsymbol{E}_x = \boldsymbol{E}_{x0}(\cos\left(\omega t - kz + \varphi_x\right)$$
$$\boldsymbol{E}_y = \boldsymbol{E}_{y0}(\cos\left(\omega t - kz + \varphi_y\right) \tag{2.31}$$
$$\boldsymbol{E} = \sqrt{\boldsymbol{E}_x^2 + \boldsymbol{E}_y^2},$$

where $\boldsymbol{E}_{x0}$ and $\boldsymbol{E}_{y0}$ are the amplitudes, and φ_x and φ_y the phases of the partial waves.

There are three possibilities for the polarization of the sum of partial waves (Figure 2.6).

a. The amplitudes of the partial waves $\boldsymbol{E}_{x0} = \boldsymbol{E}_{y0}$ are equal, and the phase shift between the two partial waves is zero $\varphi_x = \varphi_y$. The maximums occur simultaneously in both partial waves. The summary wave keeps its linear polarization (Figure 2.6a).
b. The amplitudes of the waves are equal $\boldsymbol{E}_{x0} = \boldsymbol{E}_{y0}$, and the phase shift between two partial waves $\varphi_x - \varphi_y = 90°$. The maximums of partial waves are time shifted to a quarter of a period. The sum of partial waves has vertical polarization in the beginning and horizontal after a quarter-period. The polarization plane rotates with the frequency of the wave. The sum of two partial waves is the circularly polarized wave, where the electric field rotates with the frequency of the wave ω (Figure 2.6b). The circularly polarized wave

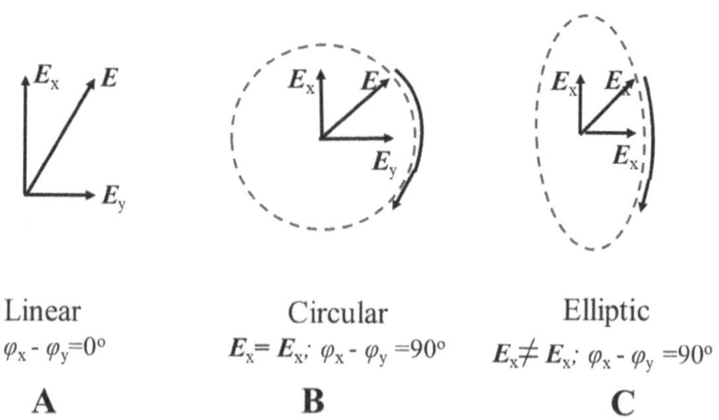

FIGURE 2.6 Linear, circular, and elliptic polarizations of electromagnetic waves.

can be divided into two linearly polarized components rotating in opposite directions.

c. The amplitudes of the waves are not equal, $\boldsymbol{E}_{x0} \neq \boldsymbol{E}_{y0}$, and the phase shift between the two partial waves is arbitrary. The phase shift causes rotation of the polarization plane. As a result, the wave is elliptically polarized (Figure 2.6c). The elliptically polarized wave can be presented as a combination of a linearly polarized wave and a circularly polarized wave.

2.6.2 The Change of Polarization upon Reflection

The electromagnetic wave emitted from the base station is linearly polarized, usually in the vertical plane, sometimes in the horizontal plane. However, the reflection from different surfaces not perpendicular to the direction of wave propagation can alter the orientation of its polarization plane and the nature of polarization. The sum of incident and reflected waves on the surface should fulfill the boundary conditions. The reflection coefficients of the horizontal (Eq. 2.24) and vertical (Eq. 2.25) components of the field vectors $\boldsymbol{E}$ and $\boldsymbol{H}$ are complex quantities that differ due to different boundary conditions (Eqs. 1.16–1.19). Therefore, the amplitude and phase of the reflected wave $\boldsymbol{E}_2$ differ from those of the incident wave $\boldsymbol{E}_1$.

$$E_1 = \cos\left(\omega t + kr + \varphi_0\right) \tag{2.32}$$

$$E_2 = E_1 R \cos\left(\omega t + kr + \varphi_0 + \varphi_R\right) \tag{2.33}$$

where R is the amplitude of the reflection coefficient and φ_R is its phase that determines the phase shift upon reflection.

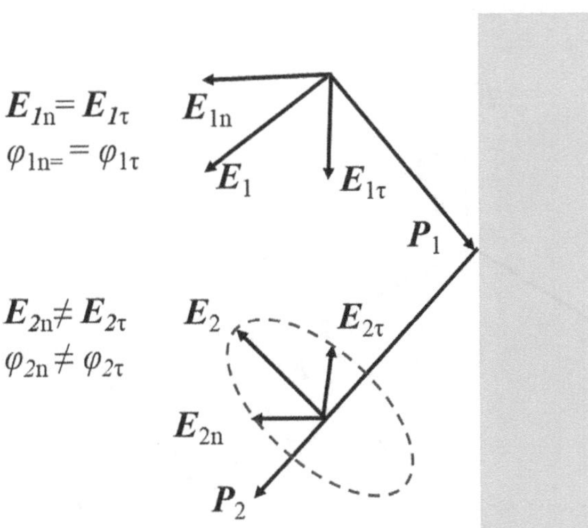

FIGURE 2.7 The linear polarization of the wave becomes elliptic upon reflection.

As a result, the polarization plane of the reflected wave is turned and the polarization of the reflected wave becomes elliptic (Figure 2.7).

The change of phase upon reflection is a constant value determined by the reflection conditions. The constant phase shift does not affect the coherence of the reflected wave.

The linear polarization of the wave emitted by a base station keeps its linear polarization only in an environment without reflecting surfaces like open fields. In cities and inside buildings, multiple reflections occur from various objects and surfaces. The electric field oscillates simultaneously in different planes, and the electromagnetic wave loses its initial linear polarization. The electromagnetic radiation in the city and inside rooms is not linearly polarized. The measurements of the field strength should be performed using sensors and antennas not sensitive to the polarization of radiation or making measurements in two orthogonal planes.

The human body and organs are of complex shapes. The incident angle of a wave propagating in the body varies in different organs and tissues. Therefore, the partial reflection (Section 2.6.2) occurs between different tissues in the body. The reflections from different tissue layers lead to an alteration of the electric field polarization.

2.7 MOVEMENT OF CHARGES IN AN ELECTROMAGNETIC WAVE

In an electromagnetic field, charged particles are simultaneously affected by the electric Lorentz's force (Eq. 1.3) and the magnetic Lorentz's force (Eq. 1.8). The electric and magnetic forces cause different movements of charged particles. The electric Lorentz's force causes direct movement in the direction of an electric field. The magnetic Lorentz's force causes orbital movement around the direction of a magnetic field with the velocity vector perpendicular to the magnetic field and the charge velocity (Figure 1.2). Perpendicular to each other, electric and magnetic fields cause a spiral-shaped movement: a charged particle moves along a spiral trajectory in the direction of wave propagation (Figure 2.8).

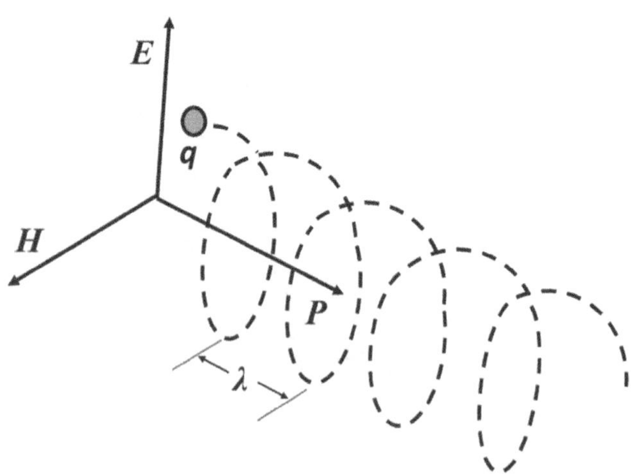

FIGURE 2.8 The trajectory of a charged particle q in an electromagnetic wave.

The phase of the wave and the sign of the charge determine whether the charge moves in the same or opposite direction with regard to the wave. Since the phase propagates with phase velocity, the charge must move at the same velocity as the phase to maintain its direction of movement. However, the charge can move only with a group velocity that is equal to the phase velocity only in an ideal medium (vacuum). In a real medium, the charge velocity is lower than the phase velocity, the direction of charge propagation alternates in time, and the charge oscillates in the electromagnetic field. Therefore, energy transfer from the wave to the charge is limited. The total impact of an electromagnetic wave on a charge, if at all, is small. In devices, like a linear accelerator, where the acceleration of particles by microwave radiation is employed, the special retardation of the phase velocity is applied.

The effect of the wave on the movement of charges in a thin channel requires an exact direction of the wave propagation toward the axes of the channel. The impact on the charge movement becomes possible only in the conditions of the retardation of the phase velocity in the channel. Due to these reasons, the electromagnetic wave can hardly directly affect the movement of ions in membrane channels.

The size of charged ions in biological tissues is much less compared to the wavelengths at radiofrequencies. Therefore, the ability of individual ions to catch the energy of radiofrequency radiation is very small. The effect can only occur as the simultaneous cumulative effect of a coherent in-phase field on a large number of ions.

The regular movement of charged particles in an electromagnetic wave can take place only in a vacuum; in a medium, the displacement of charged particles is limited by the space between the particles. In a medium, the movement is disturbed by collisions with other particles. Due to collisions, the regular movement of charges becomes irregular, random movement equivalent to thermal motion. The energy of an electromagnetic wave transfers to thermal energy.

In the case of dipolar molecules, one more kind of movement can take place in addition to the direct and orbital ones – rotational. The electric field rotates the dipolar molecule around its axis without displacing it in space. The rotational movement has no limitations due to collisions with other particles. The orientational movement of dipolar molecules can have an advantage over the displacement movement.

The character of charged particles' movement determines the character of electric permittivity and magnetic permeability of the medium.

2.8 COHERENCE OF ELECTROMAGNETIC RADIATION

What is coherence? In the first approach, I would say that this is the property of a wave to keep its phase. However, it is not precise or unambiguous, because the phase of the coherent wave also changes. The change of the phase of a propagating wave is determined by the term kz in the solution of the wave equation (2.5). The solution of the wave equation does not contain an initial phase. The initial phase is determined by the oscillations in the source of radiation, assumed to be constant, and does not impact the wave properties. However, it is important that the initial phase has a fixed magnitude but doesn't change randomly. An electromagnetic wave with a fixed phase is coherent. The phase of a coherent wave can change by a fixed amount (e.g., upon reflections) but not vary randomly. The phase of noncoherent radiation is randomly distributed in time. The concept of coherence is sometimes confusing; the correct understanding benefits the understanding of phenomena related to electromagnetic radiation.

In a linear medium, due to the linearity of the wave equation, coherent waves interfere. The summary wave is a sum of one or more separate waves (Figure 2.9). Due to the interference between the waves, minimums and maximums appear in the envelope of the summary wave.

The sum of two waves at frequencies ω_1 and ω_2, wave numbers k_1 and k_2, with equal amplitudes

$$E = E_0\left(\cos(\omega_1 t - k_1 r\right) + E_0\left(\cos\omega_2 t - k_2 r\right)$$
$$= 2E_0\cos\frac{\left(\omega_1 - \omega_2\right)t(k_1 - k_2)r}{2}\cos\frac{(\omega_1 + \omega_2)t\left(k_1 + k_2 r\right)}{2} \quad (2.34)$$

The first factor in the formula (2.30) presents the low-frequency $(\omega_1 - \omega_2)/2$ envelope of the high-frequency $(\omega_1 + \omega_2)/2$ oscillations. The electric field from three sources has a more complicated structure. The in-phase condition between the waves to ensure the maximum summary amplitude can be less frequently fulfilled for larger numbers of summarizing waves. Despite the decreasing regularity of the summary wave, the sum of coherent waves retains its coherent nature in a linear medium.

In the case of noncoherent radiation, from the point of view of electromagnetic waves, the phase is a random function of time. The random fluctuations in time make

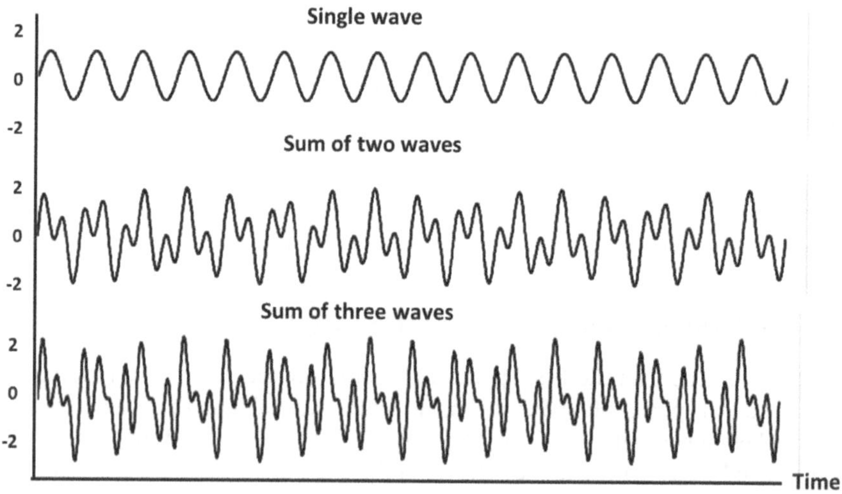

FIGURE 2.9 The graphs of electric field distribution in the time domain at r = const for a single wave calculated at f_1 = 800 MHz; the sum of two waves calculated at f_1 = 800 MHz and f_2 = 1800 MHz, and the sum of three waves calculated at frequencies f_1=800 MHz, f_2 = 1800, and f_3 = 2600 MHz.

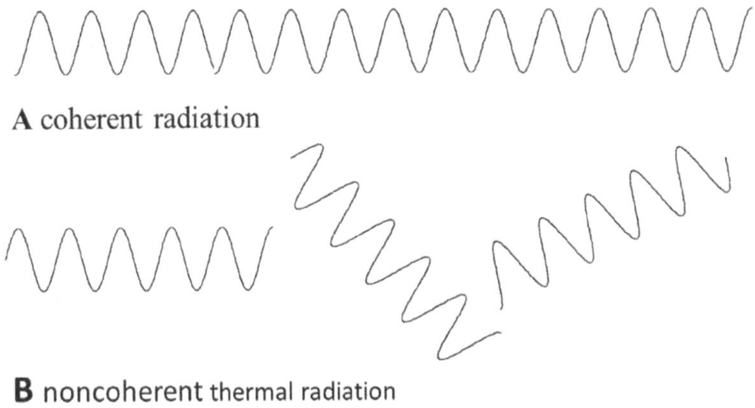

FIGURE 2.10 Illustrative picture of coherent radiation (A) and thermal radiation (B).

it impossible to describe the radiation as a regular periodic process or their combination. In the case of a nonlimited electromagnetic spectrum, the noncoherent radiation results in white noise. This is thermal radiation. It is important to underline that the nature of noncoherent thermal radiation, like the radiation from the Sun, differs from coherent radiation.

The difference between coherent and noncoherent radiations can be most easily demonstrated by comparing laser radiation with thermal radiation generated by electric lamps or the Sun (Figure 2.10). Laser radiation, following the termini, is a light amplified by the stimulated emission of radiation. In a proper active medium, the stimulating photon with a quantum of energy hf (h is Planck's constant) stimulates the generation of new photons in the phase with the simulating one (Figure 2.10A). The next, stimulated photon, in turn, generates the third in-phase photon and so on. As a result, all the photons are in-phase linked, and the laser radiation is coherent. In radiotechnical generators of harmonic oscillation, the positive feedback is also of crucial importance to ensure coherence.

In the case of thermal radiation, the photons of the radiation are not generated by the stimulation of other photons but independently by an external source of (usually thermal) energy (Figure 2.10 B). Each body with a temperature higher than 0 K generates electromagnetic radiation. The radiation from the Sun can be considered as black body radiation with a temperature of 5770 K.

2.9 SPATIAL DISTRIBUTION OF RADIATION POWER DENSITY: DIRECTIVITY OF ANTENNAS

2.9.1 Directivity of Linear Antennas

Electromagnetic energy is directed from the transmitter to the receiver using free propagation and antennas. The antenna determines the spatial distribution of radiation power (Shelkunoff, 1952; Huang and Boyle, 2008). An isotropic antenna produces a homogeneous distribution of radiation power in all directions (Figure 2.11).

The power P of the transmitter is distributed uniformly over a sphere of radius r, and the power density p of the radiation is easy to calculate

$$p = \frac{P}{4\pi r^2} \tag{2.35}$$

An elementary antenna is an electric dipole: a straight wire section shorter than the wavelength with a uniform electric current density distribution. To ensure uniform current distribution on a wire much shorter than the wavelength, capacitances must be placed at its ends to prevent the current from becoming zero. This makes the antenna like a dipole.

In the plane of dipole axes, the radiation pattern of an electric dipole looks like a horizontal number eight with the radiation maximum perpendicular to the axes (Figure 2.12). In the plane perpendicular to the axes, the radiation is directed equally in all directions, and the dipole behaves as an isotropic antenna.

Capacitances at the ends of the dipole are there to meet the current boundary conditions. Without these, the current at the ends should be zero. Whereas the length of a dipole $l \ll \lambda$, the current becomes zero throughout the dipole. To avoid the problem, traditionally a half-wave dipole (Figure 2.13) is used where the boundary conditions are met.

An electric antenna creates a linearly polarized electric field $\boldsymbol{E}$ parallel to its axes and a magnetic field perpendicular to that. The Pointing vector $\boldsymbol{P}$, perpendicular to the electric and magnetic fields, is perpendicular to the axes of the antenna. The maximum of radiation is directed perpendicularly to the antenna axes. An elementary electric antenna does not radiate in the direction of its axes. The property of the antenna not to generate radiation in the direction of its axis also applies to large

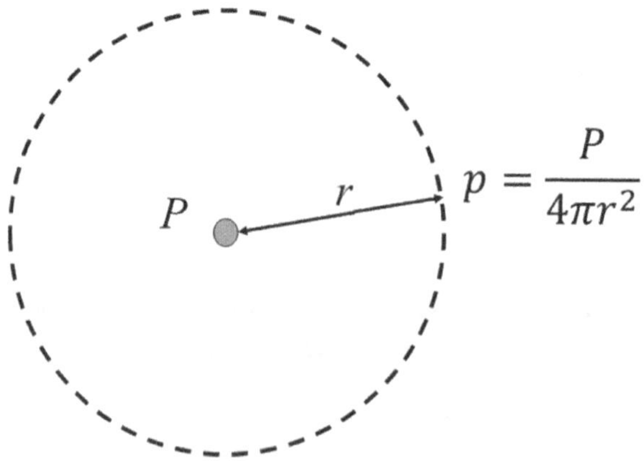

FIGURE 2.11 Power density distribution of an isotropic point source of radiation.

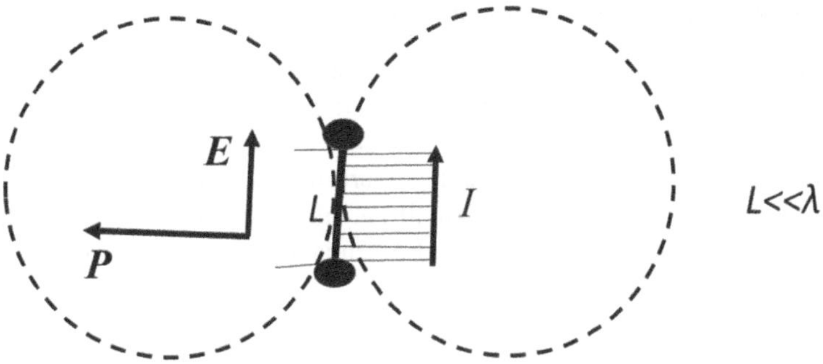

FIGURE 2.12 The directional diagram of an elementary electric dipole.

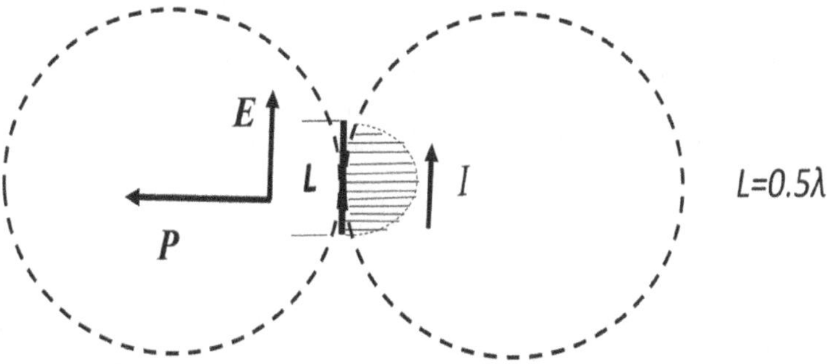

FIGURE 2.13 The directional diagram of a half-wave antenna.

antennas. From here comes the critical knowledge that no antenna radiates in the direction of its axis, straight down.

The electromagnetic field produced by the antenna has the near zone and the far zone. In the near zone $r < \lambda$, the field is reactive, the electric and magnetic components have a 90° phase shift, and the field decreases with distance r as $1/r^2$. The far zone $r > \lambda$ is the radiation zone, the electric and magnetic fields are of the same phase, and the field decreases with distance as $1/r$.

The directivity of a real antenna is the ability of the antenna to increase the power density of radiation in a predetermined direction. Directivity D is determined as a ratio of power density radiated by a real antenna to a certain direction $p(\theta, \varphi)$ to that radiated by an isotropic antenna in the same direction p_{iso} determined by zenith angle θ and azimuth angle φ regarding the axes of the antenna

$$D(\theta,\varphi) = \frac{p(\theta, \varphi)}{p_{iso}} = \frac{4\pi p(\theta, \varphi)}{P_{tot}} \tag{2.36}$$

The elementary antenna has a much shorter wavelength, the diffraction angle is large, and the directivity is low, $D = 1.5$. Frequently used, a half-wave antenna ($l = \lambda/2$) has somewhat higher directivity, $D = 1.64$. The quarter- and half-wave electric antennas are frequently used as the elements of large antenna arrays.

2.9.2 Directivity of Aperture and Array Antennas

The ability of an antenna to concentrate radiation energy in a predetermined direction has physical limitations by diffraction. The diffraction angle is determined by the ratio of wavelength to antenna size λ/l. To ensure a sharp beam and high directivity, the antenna must be large. Therefore, big aperture antennas are used. A large aperture can be replaced by an array of small antennas. The directivity of an aperture antenna can be calculated by the ratio of antenna area S to the equivalent area of a wavelength λ^2 of the radiation

$$D(\theta,\varphi) = 4\pi \frac{S}{\lambda^2} F(a,\varphi) \tag{2.37}$$

where the function $F(a,\varphi)$ describes the distribution of amplitude and phase on the antenna's aperture.

Receiving the radiation, the antenna can collect not all the energy emitted by the transmitter but only a part of that. The parameter indicating the effectiveness of the antenna in collecting radiation energy is the antenna's effective area

$$A_{ef} = \frac{\lambda^2}{4\pi} D \tag{2.38}$$

The antenna effective area A_{ef} depends on the directivity of the receiving antenna determined by the relation of the geometrical size of the antenna to the wavelength of radiation (Eq. 2.33). The power on the receiving antenna P_r depends on the directivity of the transmitting antenna as well as on the effective area of the receiving antenna. Considering the relationship (Eq. 2.35), the formula for the power on the receiving antenna can be written as

$$P_r = \frac{\lambda^2}{(4\pi r)^2} D_t D_r P_t \tag{2.39}$$

where D_t is the directivity of the transmitting antenna, D_r is the directivity of the receiving antenna, and P_t is the output power of the transmitting antenna.

The directional diagram of a large antenna is formed as a result of mutual interference of the waves from different elementary subareas of the antenna or small antennas in the array (Figure 2.13). The directional characteristics of an elementary antenna used as the elements of a large antenna are also decisive in designing the directional characteristics of the large antenna. The electromagnetic waves from different elementary antenna elements interfere with alternating phase shift $k\Delta r$, where Δr is the difference between geometrical distances from different antenna elements (Figure 2.14). The directions from two antenna elements to a point far from the antenna are parallel. Maximums form in the directions of the in-phase condition for joining waves. The number of maximums in the directional diagram n in the vertical or horizontal plane is equal to the number of half-waves on the antenna size in the corresponding plane $n = 2l/\lambda$.

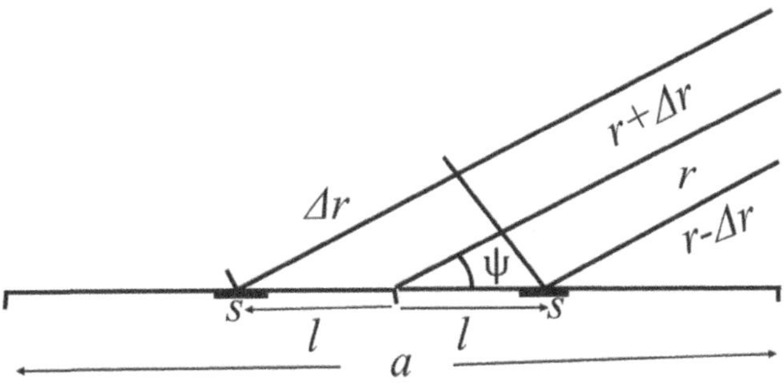

FIGURE 2.14 The interference between the radiations from antenna elements.

The phase shift between the radiations from two antenna elements s located at distance l symmetric regarding the center of the area is $\Delta r = kl \cos \psi$. The summary field E produced by the antenna area is determined as

$$E = 2\int_0^{a/2} E(s) D_s(s)\cos(ks\cos\psi)\,ds \tag{2.40}$$

where $E(s)$ is the amplitude distribution on the area and D_s is the directivity of an antenna element. In the case of an isotropic antenna element with a uniform distribution of amplitude $E_s = E_0$, the field strength produced by the area is

$$E = E_0 \frac{\sin\left(\frac{\pi a}{\lambda}\cos\psi\right)}{\frac{\pi a}{\lambda}\cos\psi} \tag{2.41}$$

The radiation picture of the antenna at uniform amplitude distribution on the aperture, described by the function $\frac{\sin x}{x}$, has one main lobe and $2a/\lambda$ lower level side lobes (Figure 2.15).

The maximum of radiation is directed perpendicularly to the antenna's aperture. An antenna does not radiate in its plane, i.e., straight down.

The maximal directivity and the level of side lobes depend on the amplitude and phase distribution in the radiating area. Homogeneous amplitude distribution ensures maximal directivity but also maximal levels of side lobes due to diffraction on antenna borders. The level of side lobes can be lowered by using the field distribution on an aperture decreasing toward the edges of the antenna. The level of side lobes can be depressed by about 40 dB compared to the main lobe.

Studies sometimes use horn antennas, where samples are placed directly into the horn. The boundary conditions (Section 1.5) on the metallic surface determine the field distribution inside the horn or any other rectangular metallic structure. Only the field components perpendicular to the wave propagation direction contribute to energy transfer (Section 1.6). The structure of a simplest wave in a waveguide

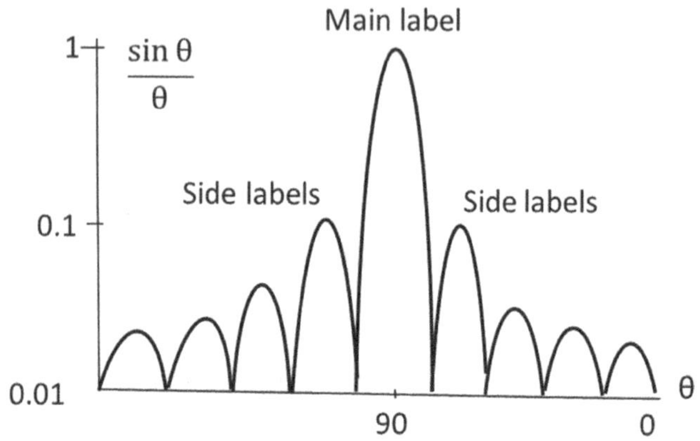

FIGURE 2.15 Directivity of an array antenna $\frac{\sin x}{x}$ at uniform amplitude distribution on the aperture.

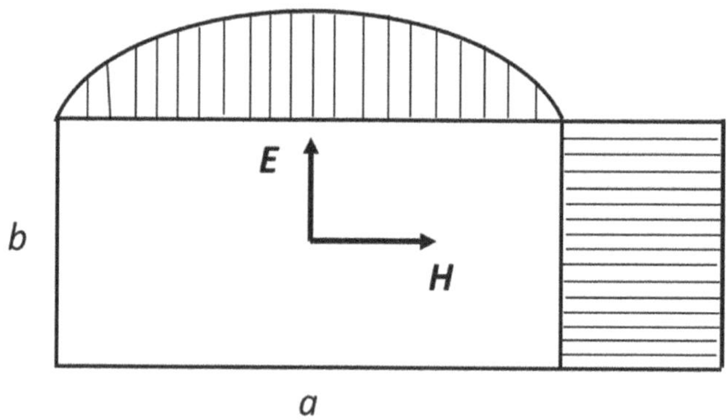

FIGURE 2.16 The distribution of electromagnetic wave vectors inside a rectangular metallic structure.

consists of one half-wave toward the wider wall *a* and a homogeneous distribution toward the shorter wall *b* (Figure 2.16).

Due to the boundary conditions inside the structure, the electric and magnetic fields should be zero at vertical walls. To provide a transfer of wave energy to a sample that should be placed in the center of the structure.

2.10 CONCLUSIONS

The following conclusions formulated from the theory of electromagnetic waves are most relevant to the impact and properties of radiation in living tissues:

- The nature of natural thermal electromagnetic radiation and technical electromagnetic radiation is different. Technical radiation used in telecommunication systems is a regular coherent radiation that can be described as an electromagnetic wave with a fixed phase and frequency. Noncoherent thermal radiation is described only by its intensity.

- Coherence of electromagnetic radiation is crucial to cause the impact of the radiation on a medium.
- Due to the linearity of field equations and dielectric polarization, the sum of two or more coherent waves in a linear medium is coherent. The waves do not interact with each other in a linear medium.
- The theory of electromagnetic waves in a lossy dielectric medium is a basis for describing the propagation and distribution of electromagnetic radiation in a living medium.
- Electromagnetic wave parameters and properties in a lossy dielectric medium differ from those in a vacuum. Wavelength in a medium is shorter, and wave phase velocity is slower compared to that in a vacuum or atmosphere.
- The depth of penetration of the electric field into a lossy dielectric medium decreases proportionally to the square root of the frequency and specific conductivity of the medium.
- The inhomogeneity of a medium causes reflection and refraction and inhomogeneous distribution of electromagnetic radiation depending on the spatial polarization of the electric field.
- The character of the initially linear polarization of the electric field can be changed and transformed into elliptic at reflection.
- The electric field component parallel to a conductive surface is zero and has its maximum at the distance of quarter-wave from the surface. The distribution of the magnetic field is opposite.
- The electric field component perpendicular to the conductive surface has its maximum on the surface and minimum at a distance of quarter-wave from the surface. The distribution of the magnetic field is opposite.
- The antenna theory describes the spatial distribution of electromagnetic radiation. The strength of the electric field decreases inversely with the distance, and the field power density decreases with the square of the distance.
- The directivities of an elementary linear antenna and an antenna array both demonstrate no radiation in the direction of antenna axes or in the plane of radiation aperture. An antenna located on the roof cannot generate strong radiation in the rooms located directly below it.

REFERENCES

Huang Y, Boyle K. (2008) *Antennas: from theory to practice*. Wiley, New York.

Marconi G, (1899). Wireless telegraphy. *Proc Inst Electr Eng*. 28(139):294.

Maxwell JC, (1873). *A treatise on electricity and magnetism*. Clarendon Press, Oxford.

Planck M, (1900). On the theory of the energy distribution law of the normal spectrum. *Verhandl, Dtsch, Phys Ges, 2. Ann. Physik*, 1:73.

Ramo S, Whinnery JR, van Duzer T, (1994). *Fields and waves in communication electronics*. Wiley, New York.

Shelkunoff SA, (1952). *Antennas: theory and practice*. Wiley, New York.

Zahn M, (1979). *Electromagnetic field theory, a problem solving approach*. John Wiley & Sons, New York.

3 Electromagnetic waves in the human body

Dielectric polarization of living tissues and wave propagation in the multilayered body

HIGHLIGHTS

- Orientational polarization of dipolar water molecules determines the electric permittivity of tissues.
- The electric permittivity and conductivity of tissues are measurable parameters.
- The electric properties of tissues depend on temperature and frequency but not on the strength of the electric field.
- The electric permittivity (related to low-level non-thermal effects) decreases, and the conductivity (related to thermal effects) increases with frequency.
- The parameters of electromagnetic waves in biological tissues differ from those in air.
- About half of the radiation power is reflected from the skin.
- Reflections on the borders between different tissues make the field distribution in the human body nonhomogeneous. Maximums of absorbed radiation power become evident on the borders between different tissues.

3.1 INTRODUCTION

The electric and magnetic forces of electromagnetic waves affect charges in the tissue structure. The electric field produces dielectric polarization of the medium (Section 1.2). The electric permittivity is a measurable parameter. There is no measurable magnetic permeability in biological tissues. The electric permittivity, describing the dielectric polarization of tissues, determines to what extent the distribution of charged particles inside tissues is affected by the external electric field.

The relative magnetic permeability of biological tissues $\mu_r = 1$. The human body behaves as a nonmagnetic dielectric medium that contains some ferromagnetic molecules, e.g., iron. However, the concentration of these molecules is small, and these do not cause a measurable impact on magnetic permeability. The mechanisms of the

DOI: 10.1201/9781003729976-4

magnetic field on human tissues are related to radical pairs (Barnes and Greenebaum, 2015). However, the mechanism of radical pairs has frequency limitations at radio-frequencies. Therefore, the electric properties of tissues determine the nature of the impact of microwave radiation on humans.

The human body is a nonhomogeneous medium containing different tissues: skin, fat, muscle, bone, brain, and other organs. The human body contains about 60% of water. Most of the water is free water contained in various body fluids: intracellular fluid, extracellular fluid, plasma, and others. Bound water is contained inside organs like the brain, liver, muscles, and others. Fat tissue contains about 10% water, and muscle tissue about 75%. Tissues are divided into high water content tissues, like muscle, brain, liver, and other organs, and low water content tissues like fat and bones. The high content of water in the tissues is a reason why the electrical properties of tissues are similar to those of water. The electric permittivity of water is about 80. The water model is relevant and frequently used to describe the dielectric polarization of human tissues.

This chapter presents data on the dielectric permittivity and conductivity of the tissues in the brain at microwave frequencies. Based on these data, the parameters of the waves and field distribution in a layered model of tissues have been estimated. This knowledge contributes to the understanding of exposure conditions and behaviors of the radiation inside the brain.

3.2 DIELECTRIC POLARIZATION OF HUMAN TISSUES

The human tissue is a nonideal dielectric with high dielectric permittivity at radio-frequencies. The polarization of tissues depends on the causing that electric field (Eq. 1.10). In a linear medium, the electric permittivity is constant and the polarization increases proportionally to the strength of the electric field. The threshold of the dielectric is not determined. The nonlinearity can occur only at very high electric fields, beginning at approximately 2–5 kV/m. When discussing external radiofrequency fields in humans, such electric field strengths are not realistic. In a linear medium, the electric permittivity of tissues is independent of the level of the electric field but depends on temperature and frequency.

The rising temperature increases the random thermal movement of particles inside tissues. The random thermal movement of particles disturbs the regular orientation of the particles in an electric field and, in this way, reduces the electric permittivity. In this way, the increase in temperature results in a decrease in electric permittivity. The temperature dependence of tissues is determined by the temperature dependence of water, which is low, about 2% per degree C.

The increased temperature also leads to an increased number of collisions between the particles. This results in an increased transfer of electromagnetic energy to thermal energy, equivalent to increased thermal losses described by the imaginary part of the electric permittivity.

The dependence on frequency, and dispersion of electric permittivity, is determined by relaxation mechanisms in tissues. Three characteristic kinds of dispersion can be distinguished (Figure 3.1).

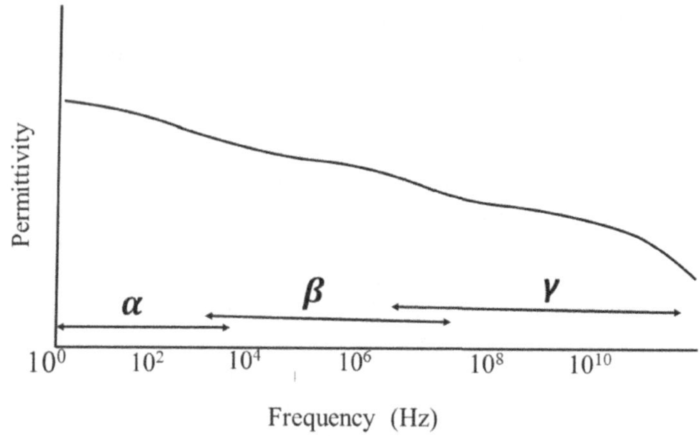

FIGURE 3.1 Three different dispersions in biological tissues.

First, the low-frequency α dispersion occurs at the hertz to kilohertz range. The α dispersion is related to ionic diffusion processes at the cellular membrane. A relaxation frequency is about 100 Hz.

Second, the β dispersion is caused by the polarization of cellular membrane and organic macromolecules. The β dispersion extends over 3–4 frequency decades centered in the hundreds of kilohertz band.

Third, the γ dispersion is related to the molecular polarization of tissue water. The γ dispersion occurs at higher frequencies in the gigahertz region. In the microwave band, the γ dispersion and the tail end of the β dispersion can be considered. Because the γ-dispersion is due to the polarization and relaxation of the water molecules, a γ-dispersion in tissues is similar to that of water (Hasted, 1973). The γ-dispersion has a relaxation frequency of 25 GHz at 310 K and persists to frequencies above 100 GHz (Foster and Schwan, 1996).

A water model (Hasted, 1973) is frequently used for describing the electric permittivity of tissues. A water molecule is an electric dipole. An external electric field rotates dipolar molecules. At radiofrequencies, the orientational polarization caused by the rotation of dipolar molecules is much higher than the polarization caused by the displacement of electrons and ions (Section 1.2).

3.3 THE DEBYE AND COLE–COLE RELAXATION MODELS OF FREQUENCY DEPENDENCE OF ELECTRIC PERMITTIVITY

The temporal response of molecules to the applied oscillating electric field follows an exponential relationship that is characterized by a decay time. The models of frequency dependence of the electric permittivity are based on relaxation equations. The first relaxation model was developed by Peter Debye (1912). The model was modified by Kenneth Stewart Cole and Robert Hugh Cole (1941) to fit better with experimental results.

The electric properties of tissues have been investigated and discussed by many researchers, Schwan (1957), Pethig (1979), Foster and Swan (1987), and others. The properties of water up to terahertz frequency have been discussed by Liebe and others in 1991.

The dependence of the electric permittivity on the frequency is caused by the inertia of particles and their relaxation time. The relaxation time must be less than the half-period of the electromagnetic field to ensure complete rotation or displacement of the particles under the influence of the alternating electric field. Due to the inertia of particles, the electric permittivity decreases with frequency. The orientational polarization, related to the rotation of dipolar molecules, dominates at the radiofrequencies up to about 300 GHz. From this frequency, the orientational polarization becomes comparable with the ionic polarization, and the electric permittivity lowers to the level of about 2. At higher frequencies, the ionic polarization becomes dominating. The electronic polarization dominates at the optical frequencies.

The imaginary part of the electric permittivity is determined by the electric losses in the tissues. The transfer of energy from the electric field to a material occurs by the collision of particles accelerated by the field with other particles. The number of collisions per unit of time is increasing with the increasing frequency. Therefore, the conductivity σ increases with frequency. As a result, the imaginary part of the electric permittivity, determined by the ratio $\sigma/\omega\varepsilon$, has a nonmonotonic dependence on the frequency with the maximum at the relaxation frequency $f\tau$.

The Debye model for dipolar polarization describes the frequency dependence of the electric permittivity:

$$\varepsilon(\omega) = \varepsilon_i + \frac{\varepsilon_i - \varepsilon_s}{1 + i\omega\tau} \tag{3.1}$$

where ε_s is the static polarization at $f = 0$, ε_i is the ionic polarization at $f > 300$ GHz, and τ is the time constant. From equation (3.1), the real part ε' and imaginary part ε'' of the electric permittivity are:

$$\varepsilon' = \varepsilon_i + \frac{\varepsilon_s - \varepsilon_i}{1 + \omega^2\tau^2} \tag{3.2}$$

$$\varepsilon'' = \frac{(\varepsilon_s - \varepsilon_i)\omega\tau}{1 + \omega^2\tau^2} \tag{3.3}$$

The Debye relaxation model has been supported by the experimental data at frequencies below 100 GHz.

The modified relaxation model, the empirical Cole–Cole model, is more precisely tuned to experimental data. The Cole–Cole equation is written as:

$$\varepsilon(\omega) = \varepsilon_i + \frac{\varepsilon_i - \varepsilon_s}{1 + (i\omega\tau)^{1-\alpha}} \tag{3.4}$$

The equations describing the frequency dependencies of real and imaginary parts of the dielectric permittivity were derived from the analysis of experimental data:

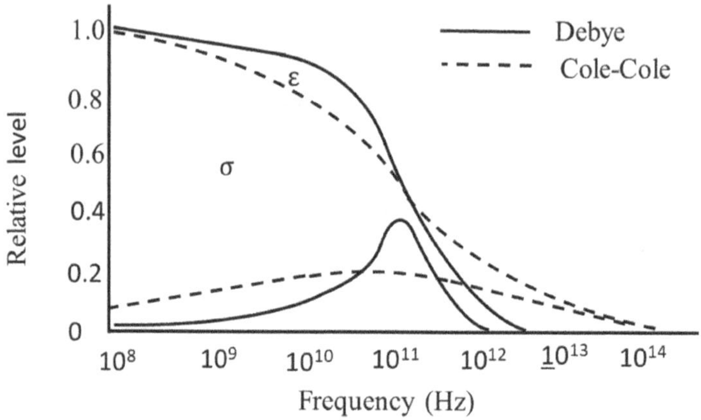

FIGURE 3.2 The relaxation curves by the Debye model (broken lines) and Cole–Cole model of frequency dependence of electric permittivity and conductivity.

$$\varepsilon' = \varepsilon_i + (\varepsilon_s - \varepsilon_i)\frac{1+(\omega t)^{1-\alpha}\sin\alpha\pi/2}{1+2(\omega t)^{1-\alpha}\sin\alpha\pi/2+(\omega t)^{2(1-\alpha)}} \tag{3.5}$$

$$\varepsilon'' = \frac{(\varepsilon_s - \varepsilon_i)(\omega\tau)^{1-\alpha}\cos\alpha\pi/2}{1+2(\omega\tau)^{1-\alpha}\sin\alpha\pi/2+(\omega\tau)^{2(1-\alpha)}} \tag{3.6}$$

Figure 3.2 presents the Debye and Cole–Cole plots for the real and imaginary parts of electric permittivity.

The developed models, supported by experimental data, predict a positive correlation between permittivity, conductivity, and tissue water content. The value of the parameter α is zero for pure water and negligible for body fluids but ≥ 0 for most tissues. In tissue, the mean relaxation time τ is generally longer than the corresponding value for pure water, indicating a restriction in the rotational ability of at least some of the tissue water molecules due to the organic environment.

The Debye and Cole–Cole relaxation models cover frequencies up to 100 GHz and are applicable to microwaves. The model of water electric properties calculated at different temperatures in the frequency range up to terahertz frequency, fitted to the experimental data, was developed by Liebe et al. (1941). The extension beyond 100 GHz was modeled by retaining the double Debye model and supplementing it with two terms by the Lorentzian resonance model describing the optical response of bound charges.

3.4 MEASURED VALUES OF ELECTRIC PERMITTIVITY AND CONDUCTIVITY OF TISSUES WITH HIGH AND LOW WATER CONTENT

The comprehensive measurements of the parameters of different tissues in a wide range from extremely low frequency to microwaves were performed by Gabriel et al. (1996). Three experimental techniques based on automatic swept-frequency network and impedance analyzers were used to measure the dielectric properties of tissue in the frequency range 10 Hz–20 GHz. The measurements were performed on 20 different animal and human tissues. The majority of measurements, except human tongue and skin in vivo, were performed using tissues after some hours (animal tissues) to some days (human tissues) after death. Further measurements of electrical parameters of various tissues in different frequency ranges were performed by Bao et al. (1997), Schmid et al. (2003), and others. The measured electric parameters of tissues reported by different scientists match well with the picture predicted by the relaxation models (Figure 3.2).

The results of measurements presented in the review by Forster and Swan (1995) and Gabriel et al. (1996a) and published by Gabriel et al. (1996b,c) and Sasaki et al. (2014), allow to get a picture of the dielectric properties of different tissues at microwave band (Figure 3.3). The graphs in Figure 3.3 show a monotonic decrease in electric permittivity with the increasing frequency. Two groups of tissues can be differentiated according to the water content: the tissues with high water content (muscle, brain, skin) and with low water content (bone, fat).

The results of measurements are in good agreement with the Cole–Cole model assuming a relation frequency of 100 GHz.

The permittivity determines the relationship between the external electric field and the field strength inside tissues that affects charges inside tissues and can cause non-thermal effects. In this way, the low-level non-thermal effect, related to permittivity, is expected to decrease at higher frequencies. The impact of conductivity on the absolute value of electric permittivity at microwaves is neglectable (Section 2.4). Due to the increase of permittivity at lower frequencies, the non-thermal effects are expected to be stronger, compared to microwaves, at electric power lines frequencies 50–60 Hz or broadcasting radiofrequency range.

The conductivity in Figure 3.3 increases with frequency. The conductivity σ determines thermal losses (Eq. 1.29) and the specific absorption rate (SAR) (Eq. 1.30). Therefore, conductivity is related to thermal effects. The increase in conductivity predicts the increase in losses and thermal effect at higher frequencies. However, the loss tangent decreases with frequency and the impact of conductivity on the absolute value of complex electric permittivity at microwaves is neglectable (Section 2.4).

The investigations of the dependence of dielectric polarization on age show a general decrease in the dielectric properties with age (Peyman et al., 2001). The trend is more apparent for brain, skull, and skin tissues and less noticeable for abdominal tissues. The alteration of the dielectric properties with age occurs due to the changes in the water content and the organic composition of tissues with age. The results would be useful in evaluating possible differences in the assessment of exposure for children and adults.

The water content decreases after death. The parameters of dead tissues are smaller than in human tissues. The difference is greater in tissues with high water content, up to 80%, and within 20% in tissues with low water content. Therefore, the post-mortem measurements show underestimated values of electric permittivity.

The alteration in water content alters the values of electric permittivity that increase in the tissues with higher water content. However, the change in water content doesn't affect the character of frequency dependence.

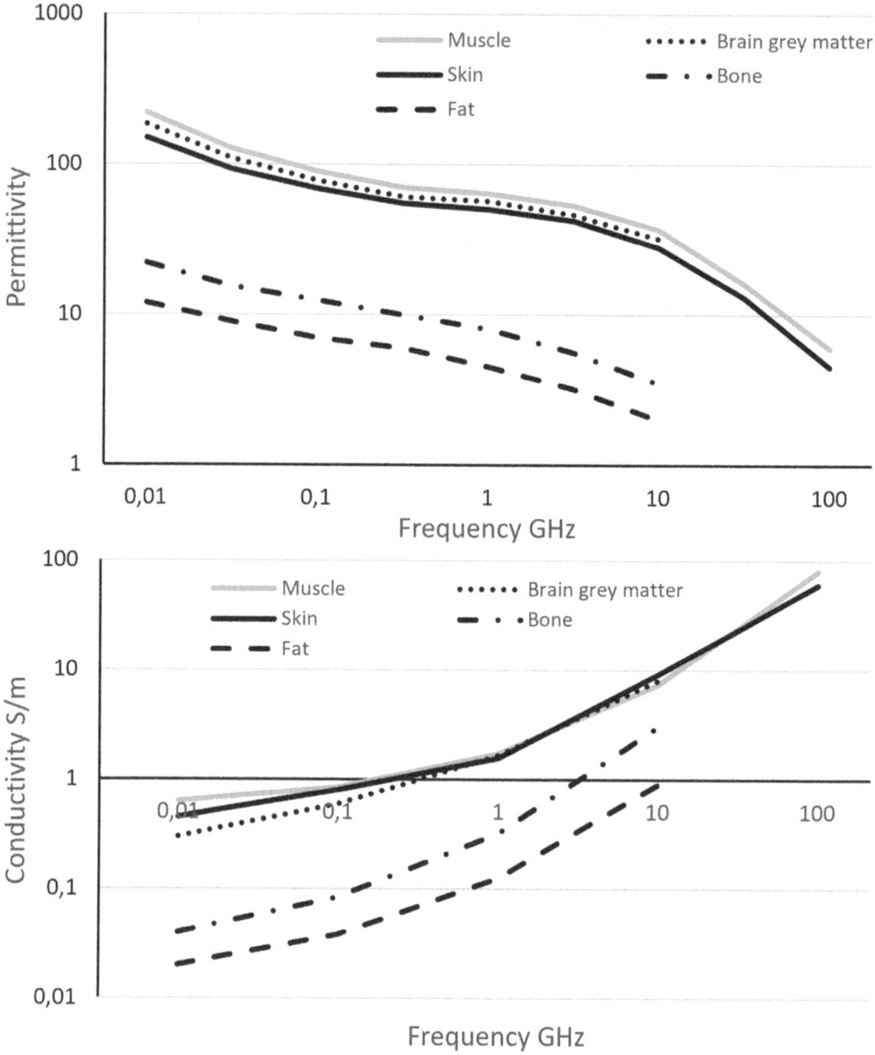

FIGURE 3.3 Frequency dependence of the electric permittivity of tissues at microwaves.

3.5 PROPERTIES OF ELECTROMAGNETIC WAVES IN A MULTILAYERED HUMAN BODY

The human body is an inhomogeneous medium with high electric permittivity and conductivity depending on the frequency of radiation. Radiofrequency radiation is reflected from the surface of the body and absorbed inside the body. Refraction and reflection of radiation occur between different layers of the body.

According to Snell's law (Eq. 2.22), due to the high permittivity of human tissues, the refraction angle, independently of the incident angle, is close to zero. The electromagnetic wave enters the body perpendicular to its surface. In electromagnetic waves propagating from air to tissue with high permittivity, Brewster's angle (Eq. 2.25) is close to 90° and not all electromagnetic energy can be transferred to the body.

The parameters of human tissues have been investigated and measured in several studies. Due to various methodologies of measurement, the reported values of the parameters are somewhat dispersed. To use electric parameters for describing the propagation of electromagnetic waves in the tissue, the singularly determined values of permittivity and conductivity at fixed frequencies should be used (Table 3.1). The wavelength inside a tissue depends not only on the frequency of radiation as in air but also on the dielectric permittivity (Eq. 2.7). Considering the values of dielectric permittivity of human tissues (Table 3.1), the wavelengths inside tissues are much shorter than those in air (Table 3.2).

A human body can be considered a lossy medium. The amplitude of the wave is decreasing in propagation inside the body (Eq. 2.17). The depth of penetration

TABLE 3.1
Relative permittivity and conductivity of muscle, brain (gray matter), skin, bone, and fat in microwave radiofrequency range

Frequency GHz	Relative permittivity				Conductivity S/m			
	Muscle	Brain gray matter	Skin	Fat	Muscle	Brain gray matter	Skin	Fat
0.01	180	289	110	12	0.63	0.37	0.8	0.02
0.1	72	90	60	6	0.89	0.69	0.9	0.03
1	54	48	50	5	1.41	1.1	1.0	0.2
10	40	40	30	3	8.3	10	10	0.9
100	6		4		80		60	

The numbers are based on data by Foster and Swan (1989), Gabriel et al. (1996), and Sasaki et al. (2014).

TABLE 3.2
The calculated wavelength, reflection coefficient, the part of reflected radiation power, and depth of penetration according to tissue parameters in Table 3.1

Frequency GHz	Wavelength		Reflection coefficient	Part of reflected radiation power %	Depth of penetration, mm
	in air	in tissue			
0.01	3000 m	286 m	0.82	67	125
0.1	300 m	38.7.m	0.77	59	33
1	30 cm	4.84 cm	0.75	56	11
10	3 cm	0.548 cm	0.69	52	3.5
100	0.3 cm	0.087 cm	0.54	30	0.7

(Table 3.2) has been calculated according to formula (2.21), considering the average values of tissue parameters in Table 3.1. The depth of penetration of the radiofrequency radiation decreases with the frequency of the radiation. To achieve deeper penetration of the electromagnetic radiation, lower frequencies must be selected.

The reflection coefficients of an electric field can be calculated according to the formula (2.26). In an inhomogeneous body, the numerical modeling of the electric field is most appropriate. The reflection from the body surface and the distribution of the electric field inside the body can be calculated based on the layered model of the body, beginning from the skin, fat, and muscle. The application of numerical methods based on the direct solution of Maxwell's equations is relevant in this case (Taflowe, 1998).

The results of calculations at different frequencies presented in Table 3.2 show that about half of the radiation power is reflected from the surface of the human body. Therefore, the special applicators are in use in devices that employ radiofrequency radiation inside the body. These applicators are matching transformers between the wave impedances in air and tissues. The part of reflected energy decreases with increasing frequency because dielectric permittivity decreases and the matching between tissue and air becomes better. The reflection coefficient has the highest value at the surface of the body where the difference in electric parameters between media is maximal.

The human body has a multilayer structure: skin, fat, muscle, etc. The reflections between the layers of different tissues inside the body are somewhat weaker compared to the reflection from the surface of the body. Reflections on the borders between the layers lead to interference between the incident and reflected waves.

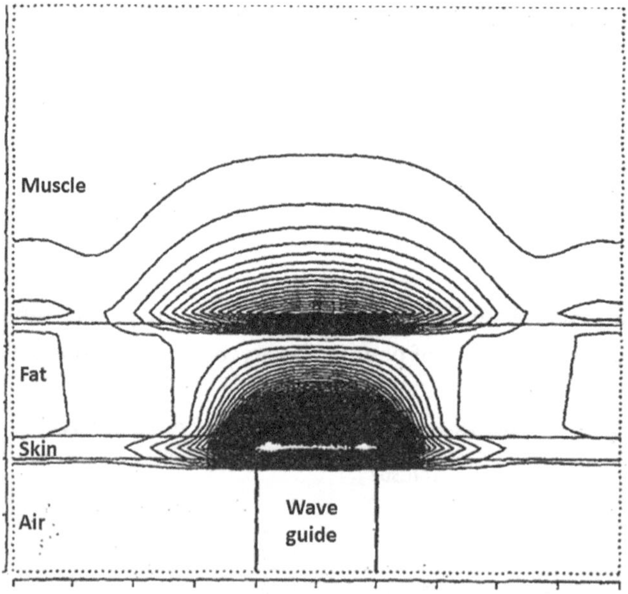

FIGURE 3.4 The distribution of absorbed power in a three-layer tissue at the frequency of 4.5 GHz.

The reflections cause remarkable spatial inhomogeneity of the wave amplitude and power distribution inside a human body. Minimums and maximums in the field, as well as in absorbed energy distribution, become evident. Figure 3.4 presents an example of numerical modeling of the distribution of absorbed power inside the three-layer body.

Diffraction at the waveguide aperture (size 0.7 wavelength) becomes evident as a broadening of the wavefront in tissues (Figure 3.4). The maximum of absorption occurs near the borders between the applicator and the skin and between the tissue layers. At the wave from the air, the reflection from the skin as well as the maximum at the skin are higher compared to matched contact between the applicator and the body. A larger part of the radiation power is expected to be absorbed by the skin.

3.6 CONCLUSIONS

Dielectric permittivity describes dielectric polarization, the primary effect of the external radiofrequency radiation in tissues caused by the radiation.

- At radiofrequencies, the orientational polarization dominates, related to the rotation of dipolar water molecules. Therefore, the electric permittivity and conductivity depend on the concentration of water in the tissue and are higher in tissues with larger parts of water.
- The electric permittivity depends on frequency and temperature but does not depend on electric field strength. Therefore, dielectric polarization is a linear process.
- The electric permittivity decreases with frequency due to the inertia of molecules. The conductivity increases due to more frequent collisions between molecules at higher frequencies.
- The dielectric permittivity of human tissues is relatively high. Therefore, the parameters of the waves inside tissues are much different from those in the atmosphere. Reflection coefficients from the human body are quite high; about half of the radiation power is reflected from the surface of the human body.
- Because of the difference in permittivity between tissues with high water content (skin, muscle, brain gray matter) and low water content (fat, brain white matter, bone), the strong reflections on the borders between different layers lead to large maximums of the radiation inside a body.

REFERENCES

Bao JZ, Lu ST, Hurt WD, (1997). Complex dielectric measurements and analysis of brain tissues in the radio and microwave frequencies. *IEEE Trans Microw Theory Tech.* 45(10):1730–1741.

Barnes FS, Greenebaum B, (2015). The effects of weak magnetic fields on radical pairs. *Bioelectromagnetics*. 2015 Jan;36(1):45–54.

Cole KS, Cole RH, (1941). Dispersion and absorption in dielectrics I. Alternating current characteristics. *J. Chem. Phys.* 9:341–351.

Foster KR, Schwan H, (1995). Dielectric properties of tissues. In: C Polk and E Postow, editors.*Handbook of biological effects of electromagnetic fields*. CRC Press, Boca Raton, FL, p. 25–102.

Gabriel C, Gabriel S, Corthout E. (1996a).The dielectric properties of biological tissues: I. Literature survey. *Phys Med Biol.* 41:2231–2249.

Gabriel S, Law RW, Gabriel C, (1996b).The dielectric properties of biological tissues: II. Measurements in the frequency range 10 Hz to 20 GHz. *Phys Med Biol.* 41:2251–2269.

Gabriel S, Lau RW, Gabriel C (1996c). The dielectric properties of biological tissues: III. Parametric models for the dielectric spectrum of tissues. *Phys. Med. Biol.* 41:2271–2271.

Hasted JB, (1973). *Aqueous dielectrics*. Chapman and Hall, London.

Liebe HJ, Hufford GA, Manabe T, (1991). A model for the complex permittivity of water at frequencies below 1 THz. *Int J Infrared Milli Waves*. 12:659–675.

Pete D, (1912). Zur Theorie der spezifischen Waerme. *Annalen der Physik*. 39(4):789–839. Pethig R, (1979). *Dielectric and electronic properties of biological materials*. John Wiley & Sons, New York.

Peyman A, Rezazadeh AA, Gabriel C, (2001). Changes in the dielectric properties of rat tissue as a function of age at microwave frequencies Phys. *Med. Biol.* 46(6):1617–1629.

Sasaki K, Wake K, Watanabe S, (2014). Measurement of the dielectric properties of the epidermis and dermis at frequencies from 0.5 GHz to 110 GHz. *Phys. Med. Biol.* 59:4739–4747.

Schwan HP, (1957). Electrical properties of tissue and cell suspensions. *Adv Biol Med Phys.* 5:147–209.

Schmid G, Neubauer G, Mazal PR, (2003). Dielectric properties of human brain tissue measured less than 10 h post-mortem at frequencies from 800 to 2450MHz. *Bioelectromagnetics*. 24:423–430.

Taflove A, (1998). *Advances in computational electrodynamics: the finite-difference time-domain method*. Artech House, Norwood, MA.

4 Bioelectromagnetic nature of the brain

From the neuronal to the system level

HIGHLIGHTS

- The neuronal processes strongly depend on temperature. It is critical to maintain temperature control in radiofrequency radiation experiments to avoid confusing the effect of radiation with the impact of temperature.
- Unlike electromagnetism, which is based on linear Maxwell's equations, the neuronal processes are described by the nonlinear Hodgkin–Huxley model.
- Living processes are nonlinear. The brain is a nonlinear system.
- The neurophysiological processes are complex and described by the theory of nonlinear dynamics, which makes the results of the impact of external electromagnetic radiation chaotic and unpredictable.
- Tissues are a linear medium concerning electromagnetic radiation. The nonlinearity becomes evident in the response of neurophysiological processes to microwave radiation but not in the properties of electromagnetic radiation in tissues determined by dielectric permittivity and conductivity.
- Due to the nonlinearity of living processes, even a minor initial disturbance can lead to significant biological effects.
- The brain's bioelectrical electroencephalographic (EEG) signals, which reflect the brain's neuronal functioning, are complex and nonlinear.
- Nonlinear signal processing methods have an advantage in processing EEG signals over linear signal processing methods.

4.1 INTRODUCTION

Electric forces play a substantial role in living systems. The electric forces hold atoms together in molecules and molecules together in cells. Bioelectromagnetic fields are produced by charges within living systems (Malmivuo and Plonsey, 1995). All physiological processes are related to the movement of charges, restructuring their configurations, and alterations in forces between them. Neuronal activity is based on the propagation of nerve pulses in nerve fibers. The external electric field is expected to affect neuronal activity.

DOI: 10.1201/9781003729976-5

The first experiments on neuromuscular effects using direct electric current were performed by Luigi Galvani (1781; Rowbottom and Susskind, 1984). He stimulated the frog's femoral nerve and demonstrated the contraction of a frog leg muscle. The first experiments with radiofrequency electric current were conducted by Jacques Arsčne d'Arsonval in 1891–1893 (Malmivuo and Plonsy, 1995). d'Arsonval demonstrated the stimulation of muscle and nerve by electric and magnetic fields. He reported a positive effect of the high-frequency electric field of about 5 MHz frequency and introduced the use of high-frequency currents in electrotherapy.

The success in physiology underlying neuronal functioning and bioelectric activity in neural systems was achieved much later, in the 1950s. John Eccles explained the synaptic transmission of action potential between neurons in the 1950s. Hodgkin and Huxley developed, based on experimental results, their model of nerve pulse propagation in 1953. The Nobel Prize in Physiology or Medicine 1963 was awarded jointly to Sir John Carew Eccles, Alan Lloyd Hodgkin, and Andrew Fielding Huxley "for their discoveries concerning the ionic mechanisms involved in excitation and inhibition in the peripheral and central portions of the nerve cell membrane."

The importance of the Hodgkin–Huxley nonlinear equation in neuronal activity is similar to that of the linear wave equations in electromagnetic wave propagation in media. The principal difference between the characters of the two equations, the linearity of the wave equation and the nonlinearity of the Hodgkin–Huxley equation, leads to a substantial difference in the properties of electrical and biological processes. Biological processes in living media are nonlinear. Electric processes in both non-living and living media are linear (Chapter 3).

The brain is a complex system. The functioning of the brain is based on the cooperation of about 10^{11} neurons connected by about 10^{14} synapses, dependent on a huge amount of parameters at different levels, and described by a nonlinear model. The outcome of the nonlinear dynamics equation is unpredictable. An adequate physical model of brain functioning is still missing. Brain neuronal activity is expressed as neuronal oscillations that can provide some information about the brain functioning. The neuronal oscillations are reflected by oscillating potentials on the scalp recorded as an encephalographic signal. This chapter presents the general principles of bioelectromagnetism and shows that nonlinear living systems behave differently from linear non-living systems. The nonlinearity of the neuronal system significantly affects the impact of microwave radiation on the brain.

4.2 THE BASICS OF THE BRAIN BIOELECTROMAGNETISM

The neurons are connected by axons and synapses (Figure 4.1). Electric nerve pulses generated on the membrane of neurons propagate through axons and pass through synapses between neurons.

A semi-transparent membrane of a neuron consists of lipids, proteins, and other biomolecules. The thickness of a membrane is about 7.5–10 nm. The ions can pass through membrane channels (Figure 4.2) from the extracellular to the intercellular medium or in the opposite direction. The comparison between the outcome of

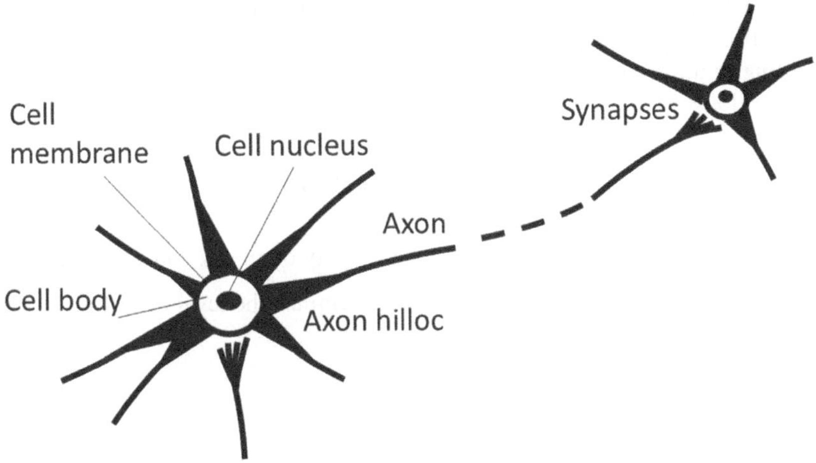

FIGURE 4.1 The structure of a neuron.

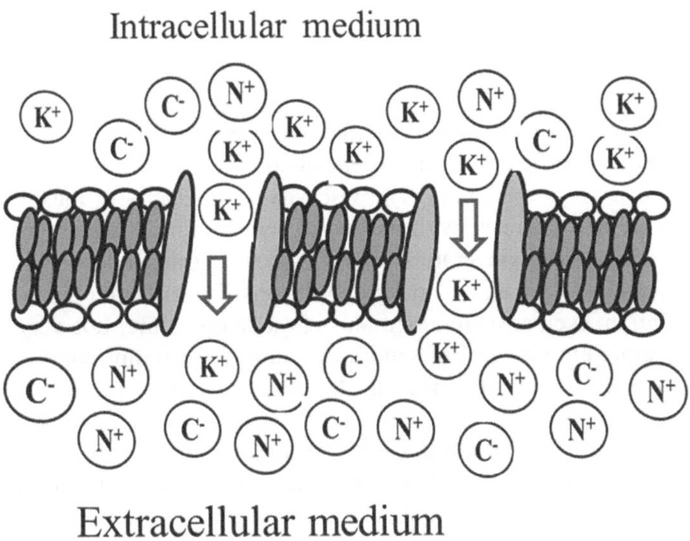

FIGURE 4.2 The excitation of the cell membrane.

computer modeling based on a nonlinear neuron model and experimental results leads to the conclusion that at least 14,000 ion channels are in a neuron (Buchholz et al., 2002).

The movement of ions through channels is caused by two factors: first, by the electric Lorentz forces (Eq. 1.3) due to different numbers of positive–negative ions in extracellular and intracellular media that produce an electric field; second, by the difference between the concentrations of the ions outside and inside the membrane.

The motion of ions depends on their velocity in extracellular and intercellular media. The membrane potential is determined by the balance of concentration according to the Nernst–Planck continuity equation describing the dynamics of ion concentration c as

$$\frac{dc}{dt} = -D\nabla c + c\boldsymbol{v} + cq\boldsymbol{E} \tag{4.1}$$

where D is the diffusion coefficient, $\boldsymbol{v}$ is the velocity of an ion, and q is the charge of an ion. The equilibrium conditions $\frac{dc}{dt} = 0$ determines the level of resting potential on the cell membrane as

$$D\nabla c = c\boldsymbol{v} + cq\boldsymbol{E} \tag{4.2}$$

The condition of a resting potential determines the balance between currents related to diffusion $D \nabla c$ and conductivity currents caused by the electric field related to membrane potential. The resting membrane potential is about 50–70 mV. The diffusion coefficient D in equation (4.1) is determined by Stokes–Einstein's equation.

$$D = \frac{kT}{\chi} \tag{4.3}$$

where χ is the friction coefficient determined by viscosity.

The diffusion coefficient strongly depends on the temperature of the medium and the viscosity of the medium. The viscosity and velocity of ions also depend on temperature and decrease with increasing temperature. These relations of the neuronal processes on temperature lead to the strong dependence of membrane potential on temperature and contribute to the possible thermal effect of radiofrequency radiation, even at small alterations of the temperature. It is very important to ensure temperature control of samples or animals in radiofrequency radiation effect experiments to avoid confusing the effect of radiation with the effect of temperature.

On the contrary, experimental studies have demonstrated the increase in the movement of particles caused by low-level microwave exposure at constant temperature (Aly et al., 2008; Hinrikus et al., 2015). These findings support the decrease of viscosity and the increase of diffusion produced by low-level microwave exposure (in detail discussed in Section 5.5).

Nerve impulses travel from one neuron to another through synapses (Figure 4.3). Activation of a presynaptic neuron results in the release of neurotransmitters into the synaptic clefts. Due to diffusion, the neurotransmitters move from the presynaptic membrane to the postsynaptic membrane and activate receptors.

The synapses of different types have either a stimulating or inhibiting effect. A nerve cell receives signals from many different synapses, and the effect is determined by which type prevails (Gerstner et al., 2014). The time delay of nerve pulses conveyed from one cell to another in a synapse was estimated from measurements of small variations in electrical charges at contact surfaces between nerve cells,

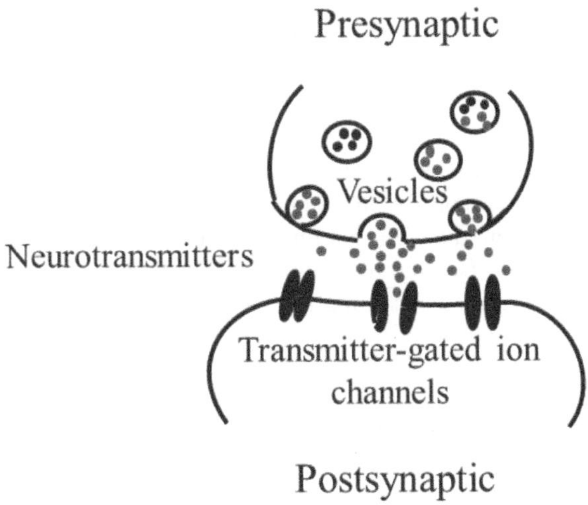

FIGURE 4.3 The scheme of synapsis.

first by John Eccles in the early 1950s. The time of passing nerve pulses in synapses varies between different types of synapses in the range of 1–15 ms (Gerstner et al., 2014).

4.3 THE EXCITATION OF NEURONS AND PROPAGATION OF ACTION POTENTIAL

The decline of the cell membrane potential from the resting one leads to the excitation of the cell. Alan Hodgkin and Andrew Huxley proposed the model of excitation of neuronal cells in 1952. In the same years, John Eccles showed how nerve impulses are conveyed from one cell to another through synapses.

The Hodgkin–Huxley model comprises basic electric circuit components such as voltage source V, resistance R, and capacity C. The model describes the total membrane current as a parallel connection of currents related to sodium ions I_{Na}, potassium ions I_K, and leakage I_l current, plus the current I_C through membrane capacity. Figure 4.3 presents the illustrative diagram of an excitable membrane channel.

The membrane current

$$I_m = C_m \frac{dV}{dt} + g_K n^4 (V - V_K) + g_{Na} m^3 h (V - V_{Na}) + g_l (V - V_l) \tag{4.2}$$

where V is the difference between the membrane potential from resting potential. The parameters n, m, and h

$$\frac{dn}{dt} = \alpha_n (1 - n) - \beta_n \tag{4.3}$$

$$\frac{dm}{dt} = \alpha_m (1 - m) - \alpha_m \tag{4.4}$$

$$\frac{dh}{dt} = \alpha_h (1-h) - \beta_h \qquad (4.5)$$

The parameters α_n, α_m, and α_h and β_n, β_m, and β_h in equations (4.3–4.5) are rate constants that depend on voltage but not on time. The parameter n represents the proportion of the particles inside the membrane, and $(1 - n)$ represents the proportion that are outside the membrane. The parameter α_n determines the rate of transfer from outside to inside, while β_n the transfer in the opposite direction. These parameters are numerically determined according to experimental data (Hodgkin and Huxley, 1952):

$$\alpha_n = 0.01(V-10) / \left(\exp^{\frac{V+10}{10}} - 1 \right) \qquad (4.6)$$

$$\alpha_m = 0.1(V-25) / \left(\exp^{\frac{V+25}{10}} - 1 \right) \qquad (4.7)$$

$$\alpha_h = 0.07 \exp(V/20) \qquad (4.8)$$

$$\beta_n = 0.125 \exp(V/80) \qquad (4.9)$$

$$\beta_m = 4 \exp(V/18) \qquad (4.10)$$

$$\beta_h = 1 \Big/ \left(\exp^{\frac{V+30}{10}} + 1 \right) \qquad (4.11)$$

All these parameters have an exponential dependence on the V.

The propagation of membrane current in a fiber is described by the equation (Hodgkin and Huxley, 1952):

$$I = \frac{a}{2R} \frac{\partial^2 V}{\partial t^2} \qquad (4.12)$$

where R is the internal resistance (resistance of the axoplasm) that indicates the ability of charges to move along the inner surface of the axon and a is the radius of the fiber. Substitution of membrane current I_m in equation (4.2) by equation (4.12) results in the wave equation of nerve pulse propagation in a neural fiber (Hodgkin and Huxley, 1952):

$$\frac{a}{2R} \frac{\partial^2 V}{\partial t^2} = C_m \frac{\partial V}{\partial t} + g_K n^4 (V - V_K) + g_{Na} m^3 h (V - V_{Na}) + g_l (V - V_l) \qquad (4.13)$$

where the coefficient $a/2R$ characterizes propagation velocity.

Equation (4.13) is a nonlinear wave equation describing the nerve pulse propagation. The parameters n, m, and h are functions of V. The analytical solution of a nonlinear dynamic equation is not possible. The nature of neuronal dynamics described by nonlinear wave equations is fundamentally different from the electrodynamics described by linear wave equations.

Unlike linear dynamics, the nonlinear dynamics equations cannot be solved analytically but only investigated using numerical methods. Nonlinear equations depend on the initial conditions and can lead to different results.

The results of the numerical investigation of equation (4.13) performed by Hodgkin and Huxley (1952) and by several other researchers later allowed us to describe the main properties of the nerve pulse and its propagation. A typical shape of a nerve pulse is presented in Figure 4.4.

The depolarization of the neuron membrane during excitation is followed by repolarization during the recovery period. The membrane potential changes from positive with respect to the resting potential during excitatory periods to negative during recovery periods. The refractory period of about 1 ms is the time taken by the membrane to get polarized again. The nerve pulse of a neuron travels along the axon to a postsynaptic neuron. The nerve pulse shape changes very little between one spike and the next. Different neuron types have different shapes and the duration of pulses, and the shape of the spike affects neurotransmitter release.

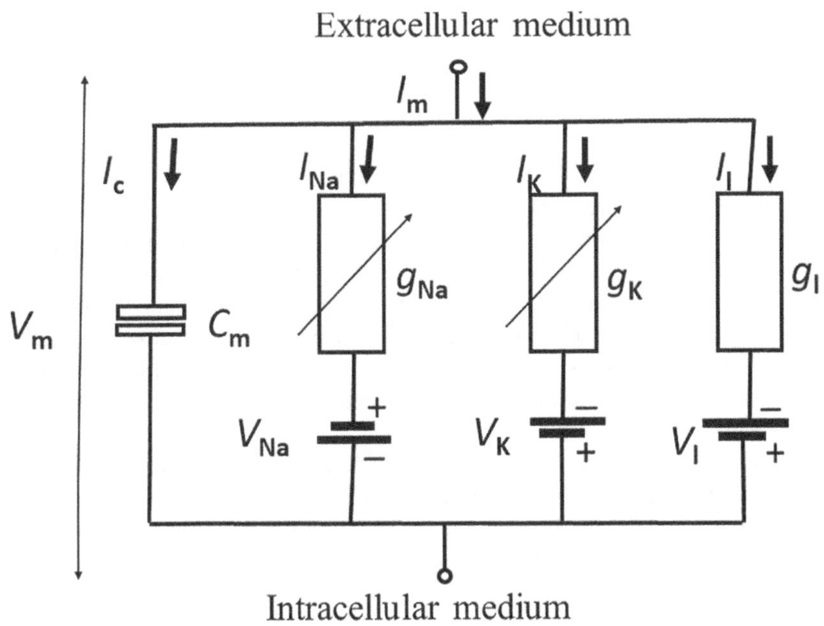

FIGURE 4.4 The electric diagram of the Hodgkin–Huxley model.

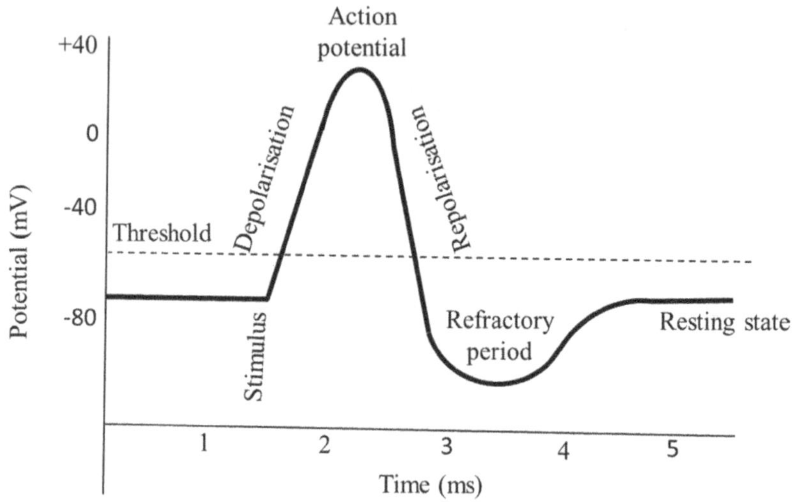

FIGURE 4.5 Nerve pulse.

The nerve pulse propagation velocity has been discussed in detail by Muler and Markin (1977) using an idealized nonlinear ionic current function. Assuming the constancy of ionic current, the velocity of the nerve pulse propagation v in an unmyelinated axon is determined as

$$v = \sqrt{\frac{a}{2RC_m}} \tag{4.14}$$

where a is the diameter of the nerve fiber, R is the resistance of the axoplasm, and C_m is membrane capacitance per unit length.

The nerve pulse propagation velocity increases with the diameter of the nerve fiber and decreases linearly with the time constant of membrane currents $\theta = \sqrt{R}C_m$.

As the velocity of nerve pulse propagation depends on the diameter of the fiber, the velocity differs between individuals and fiber types (motor, sensory, autonomic). The nerve pulse velocities vary from 120 to 0.5 m/s. These values are much lower compared to the propagation velocity of electromagnetic waves in brain tissues, about 4×10^7 m/s.

The functioning of the neuronal system is described by nonlinear equations, and their properties depend on many variables (Eqs. 4.2–4.11). These variables, in turn, depend on other variables. The complexity of living systems leads to their nonlinear nature, different from that of linear systems.

4.4 BIOELECTRIC OSCILLATIONS IN THE BRAIN

The cycle of a neuron excitation (Figure 4.4) consists of excitatory and inhibitory segments where the action potential is higher or lower than the resting potential in an undisturbed neuron. These segments describe the processes of polarization and repolarization of the neuron membrane, respectively. The neuronal activity has an oscillating nature. The oscillations occur from single neurons to neuronal populations and brain areas.

Oscillations are a prominent feature of neuronal activity, and the synchronization of the oscillations is a likely mechanism for cerebral integration for neural communication (Buzsáki and Draguhn, 2004; Schnitzler and Gross, 2005). Functional connectivity between oscillations in various brain areas is crucial in brain functioning. The relationships between large-scale and small-scale functional connectivity are a key issue in the brain neural network architecture reflecting physiology at various levels, from molecules to organs and systems that are affected by brain disorders (Bartsch et al., 2015; Basset et al., 2006; Fingelkurts et al., 2007; Lin et al., 2016; Päeske et al., 2020).

The properties of neuronal oscillators are the result of the physical architecture of neuronal networks and the limited speed of neuronal communication due to axon conduction and synaptic delays (Buzsáki, 2006; Nunez, 1995). Higher-frequency oscillations originate from a smaller neuronal population, as low-frequency oscillations encompass larger populations. The frequency of single neuron oscillations is about 1000–2000 Hz. The spectrum of brain oscillations covers the frequency range from single neuron spiking to long-range communication between different neural assemblies at lower frequencies.

The unperturbed brain is a complex system of numerous self-governed oscillations, but the content of these rhythms is still poorly understood (Steriade and Timofeev, 2003). However, some results reported in various studies by several researchers have demonstrated that the brain's physiological oscillators act very similarly to electric oscillators. Brain oscillators, like most biological rhythms, belong to limit-cycle and weakly chaotic oscillators and share features of both harmonic and relaxation oscillators (Buzsáki and Draguhn, 2004).

Detailed biophysical studies revealed that even single neurons are endowed with complex dynamics, including their intrinsic abilities to resonate and oscillate at multiple frequencies (Hutcheon and Yarom, 2000). The spectrum of brain oscillations covers the frequency range from single neurons spiking to long-range communication between different neural assemblies (Buzsáki and Draguhn, 2004; Schnitzler and Gross, 2005). The augmenting properties of resonators–oscillators are also active at the network level, and coherent summation of oscillators is an effective mechanism for the detection and amplification of weak signals (Buzsáki and Draguhn, 2004; Steriade and Timofeev, 2003). The ability of neuronal assemblies to synchronize depends on the coupling strength and the distribution of natural frequencies (Buzsáki and Draguhn, 2004).

The neuronal oscillations contain information about brain functioning, the state of the brain, and brain health. Any external factor affecting the brain causes alterations in neuronal oscillations. The nonlinearity of the brain response to external stressors makes the neural system very sensitive to microwave radiation. This is the reason why the microwave effects in human EEG are reported in many studies by independent groups (Section 6.2).

4.5 ELECTROENCEPHALOGRAPHIC SIGNALS – AS AN INDICATOR OF BRAIN FUNCTIONING

The oscillating bioelectric fields consist of information about the state of the brain. A patient-friendly, simple electroencephalographic (EEG) method employs the recording of the oscillations on the scalp. The EEG method is non-invasive and cost-effective and is therefore widely used for getting information about brain functioning. The first EEG recording was made by Hans Berger who also introduced the term electroencephalogram (Berger and Gray, 1929).

The EEG method employs the recording of electric signals from electrodes placed at fixed points on the subject's head. The positions electrodes follow the basic international 10–20 electrode placement system with 21 electrodes (Figure 4.6).

The modified electrode placement systems include 32 or 64 electrodes. The additional electrodes are placed between the electrodes of the basic 10–20 system.

Reflecting neuronal activity of the brain in real-time, the EEG signal consists of the contribution from many simultaneously oscillating neural areas and neuronal processes that make the signal nonregular and complex. The EEG signal has no characteristic shape, fluctuates, and has high natural variability (Figure 4.8). The

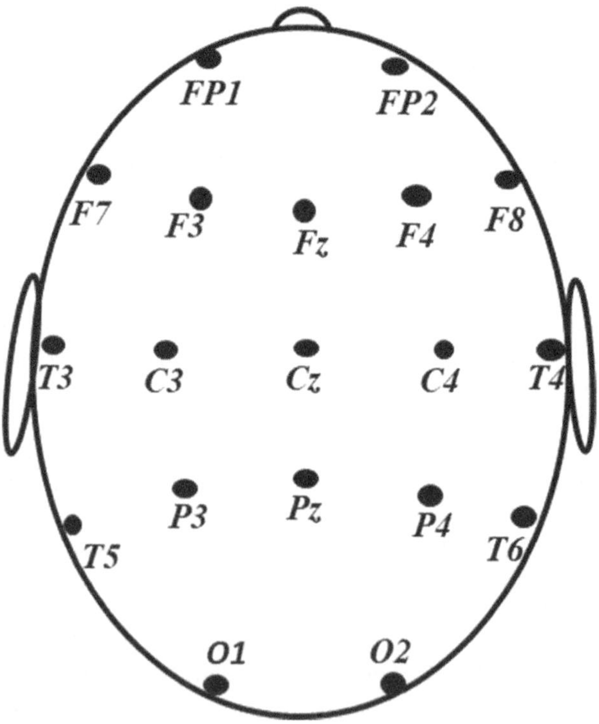

FIGURE 4.6 The international 10–20 system of placement of electrodes.

EEG signal is nonstationary, and its statistical characteristics change over time. Due to nonlinearity, nonstationarity, and a high level of fluctuations, the EEG signals are difficult to interpret. Signal processing is required to interpret the EEG picture (Figure 4.7).

The simplest and most easily applicable method of EEG signal processing is spectral analysis. However, the widely used signal processing methods of spectral analysis presume the stationarity of the signal. The brain bioelectric signals are generated by a simultaneous activity of multiple sources modulated by different physiological factors, which are intermittent and nonstationary by their nature. Therefore, the EEG signal is expected to be intermittent and nonstationary. In neurological practice, the EEG signal is considered quasi-stationary for some seconds. The short-term quasi-stationarity allows the application of spectral analyses

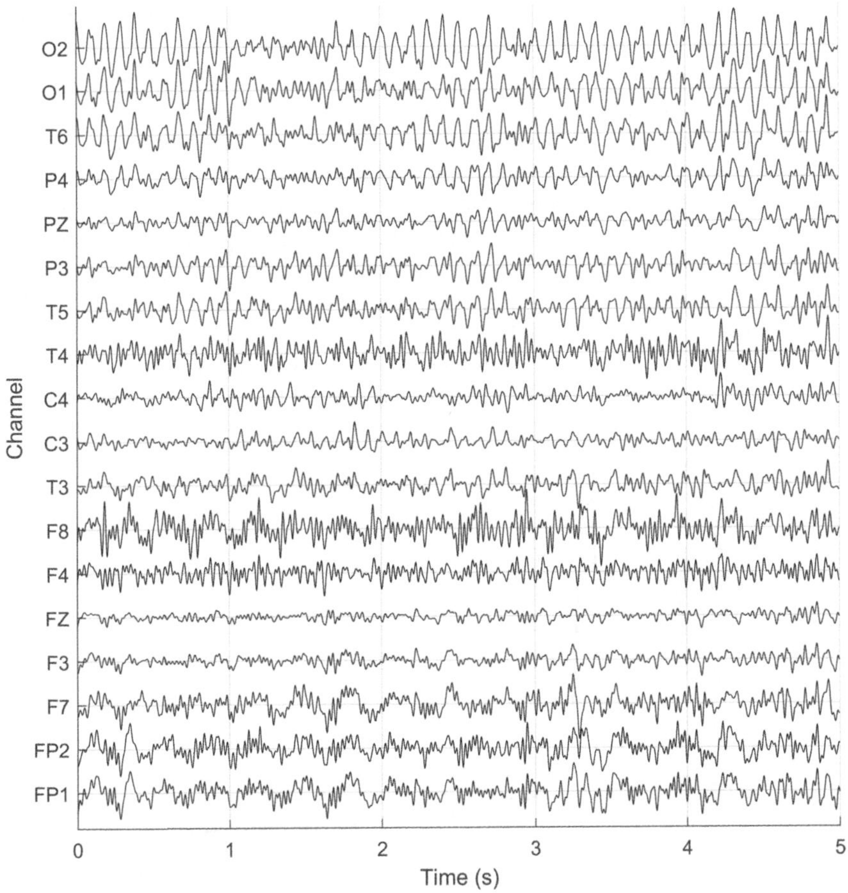

FIGURE 4.7 Raw EEG signals recorded from electrodes.

in the EEG signal processing. The calculated short signal segment values of a parameter can be averaged throughout the recording, getting a picture of the signal (Figure 4.8).

The distribution of the EEG power spectral density over the EEG spectrum reflects the state of the brain. The EEG signal power calculated within the determined frequency bands of the EEG spectrum can be used in further analyses. Traditionally used EEG frequency bands are: delta 0.5–3 Hz, theta 4–7 Hz, alpha 8–12 Hz, beta 13–30 Hz, and gamma higher than 30 Hz. The EEG frequency bands are known to be associated with specific neurophysiological and cognitive processes. The lower EEG frequencies are associated with sleep, the delta band with deep sleep, and theta with memory and drowsiness. Alpha is related to arousal, relaxation, and attention control. The beta band is associated with active thinking and motor planning. The highest frequency band, gamma, reflects higher-order cognitive functions such as perception and consciousness.

The spectrum of eyes-closed resting EEG spectrum has a specific shape with a high alpha maximum (Figure 4.8) that decreases in the eye-closed EEG. The power spectral density decreases at frequencies higher alpha peak. The gamma band higher than 30 Hz contains only about 8% of the total EEG power.

More complicated is the detection of the very short brain response to a cognitive stimulus, event-related evoked potential, as the time averaging of a short signal is impossible. For the detection of event-related potential, a special method is in use: the stimuli are repeated many times (up to hundreds); the responses are synchronized with the stimuli and averaged. The event-related potentials have a specific shape with characteristic positive and negative peaks delayed with respect to the stimuli by hundreds of milliseconds (Figure 4.9).

In clinical practice, the EEG frequency band power markers are in use. The EEG power markers have also been successfully used by many authors in detecting

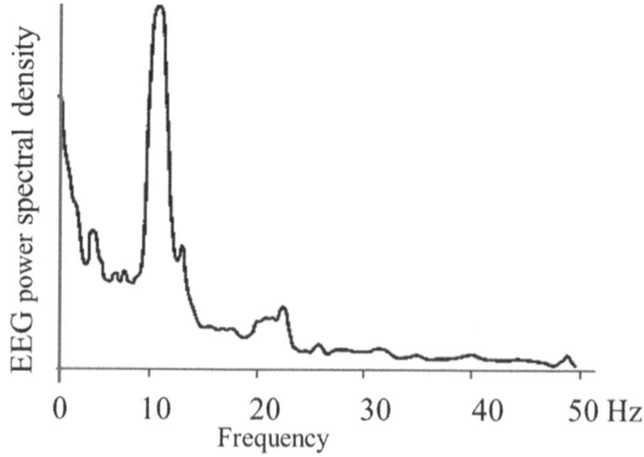

FIGURE 4.8 The spectrum of the EEG signal.

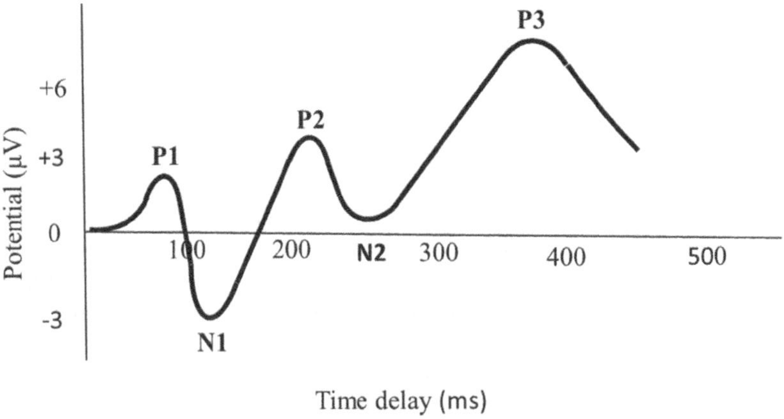

FIGURE 4.9 A visual event-related potential.

radiofrequency radiation-induced changes in EEG (Croft et al. 2008, 2010; Curcio et al., 2015; Hinrikus et al., 2008a, 2008b; Lustenberger et al., 2015; Veccio et al., 2010). However, these alterations are small, and achieving a statistically significant outcome isn't easy.

Since the brain is a nonlinear system, EEG signals are also nonlinear. Nonlinear measures have been devised to reflect the nonstationarity in the best possible way. Therefore, nonlinear methods can enable us to achieve more information about the nature of the EEG compared to linear ones. The nonlinear method can be more appropriate in EEG signal processing and is expected to have an advantage over the linear band power method. During the last decades, nonlinear EEG markers like fractal dimension, entropy, and others have been widely used in EEG processing to detect mental disorders like depression, anxiety, and others (Ahmadlou et al., 2012; Bachmann et al., 2018; Hosseinifard et al., 2013). The correlation existing between linear and nonlinear markers shows that different EEG features and markers, reflecting brain functioning, provide some overlapping information (Päeske et al., 2023). In the comparison between markers of different natures, Higuchi's fractal dimension has some advantages being correlated with the majority (87%) of other markers and in this way reflecting a wide spectrum of EEG features.

The advantage of the Higuchi fractal dimension over the power method in the detection of the microwave effect in the brain has been demonstrated in the experiments with low-level 450 MHz radiation modulated at 40 Hz. The statistical difference between exposed and sham-exposed ten segments of the signal, detected using Higuchi's fractal dimension, was much higher, $p = 0.007$, compared to that using alpha band power, where $p = 0.04$ for the same segments.

Only a few researchers have used nonlinear markers in the detection of microwave-related changes in EEG. The application of nonlinear EEG markers for the detection of microwave effects is a promising direction for further investigations.

4.6 LINEAR AND NONLINEAR PROPERTIES OF THE NEURONAL SYSTEM

4.6.1 Linear Dynamics

Linearity is the simplest, straight relationship between the input and output, the initial stressor, and the result in an affected system or processes. The conditions of the propagation of electromagnetic waves in the human body are determined by the electric permittivity and conductivity of tissues. The constant value of the permittivity at a fixed frequency allows us to consider the human tissues as a linear medium. A linear dynamics equation (4.15) describes the temporal dynamics of the system or process, where the output is proportional to the input

$$\frac{dA(t)}{dt} = aA(t) \tag{4.15}$$

Maxwell equations, which describe properties of electromagnetic fields and waves, are linear. The equation derived from the wave equation describing the propagation of electromagnetic waves in a linear medium is a linear equation.

The properties of electromagnetic radiation in tissues are determined by a single parameter, electric permittivity. The linearity assumes a constant value of the electric permittivity independently of the level of field strengths. The constancy of electric permittivity with the level of field strength is approved by the current knowledge of electromagnetism. The living tissues behave as a linear medium regarding electromagnetic waves.

Linear differential equations can be solved analytically. A linear differential equation has only one possible solution for fixed initial and boundary conditions. A change in the initial or boundary conditions in the wave equation can shift the wave in time or space but results in an identical outcome and does not change the character of the process. The linear theory of electrodynamics, based on the linearity of systems and media, is widely used in microwave devices, antennas, and wave propagation. The dielectric polarization is linear in a linear medium (Eq. 1.10).

4.6.2 Nonlinear Dynamics

Living systems are dynamic and able to adapt to alternating environments and develop and renew their components. Unlike non-living material, living objects are in permanent motion and undergo alterations at different levels from cellular to system. The biological systems are nonlinear in the electromagnetic field (Adey and Lawrence, 1984; Barnes and Hu, 1980).

The brain neuronal activity and functioning are based on the movements of ions through neuron membranes to the propagation of nerve pulses in neural networks. The brain is a highly complex system containing 10^{11} neurons oscillating at frequencies up to 10^3 Hz and connected through about 10^{14} synapses. The nonlinear Hodgkin–Huxley equations describe the neurophysiologic processes in the brain. These processes depend on many variables. These variables, in turn, depend on other variables at different levels of the brain structure. The complexity of the neurophysiologic processes in the human brain leads to nonlinear dynamics and chaos.

Nonlinearity becomes evident as a complicated, non-straight relationship between the input and output, the initial stressor, and the result in an affected system or processes (Kaplan and Glass, 1995; Strogatz, 2015; Thompson and Stewart, 2002). A nonlinear dynamic equation (4.16) describes the temporal dynamics of the system or process, where the output is correlated but not proportional to the input. The nonlinear term in dynamic equations reflects the complexity of systems and processes, as

$$\frac{dA(t)}{dt} = aA(t) + A^2(t) + \ldots. \tag{4.16}$$

Nonlinear dynamics equations cannot be solved analytically. The outcome can only be investigated using computer modeling. The character of the outcome depends on the initial conditions. The nonlinear dynamics results in fractality, bifurcation, and other nonlinear phenomena (Izhikevich, 2007). In a nonlinear system, even a very small change in initial conditions can cause the system behave completely differently. The result is unpredictable and looks random. However, the process is not random but deterministic chaos. The nonlinearity, e.g., in some nonlinear crystals, leads to the spectrum components at combination frequencies at multiple frequency radiation, and to the generation of harmonics in microwaves and optics.

Edward Norton Lorenz first described deterministic chaos in meteorology in 1963. Lorenz found that the solutions to nonlinear dynamic equations are nonperiodic and chaotic. Slightly different initial conditions would lead to a totally different state of the system after some time. He demonstrated that a very small change can trigger a very large process far from it and much later. Lorenz also introduced the concept of deterministic chaos and demonstrated that there is a structure in the chaos. The Lorenz equations are a system of three ordinary differential equations (Figure 4.10):

$$\frac{dx}{dt} = \alpha(y - x) \tag{4.17}$$

$$\frac{dy}{dt} = x(p - z) - y \tag{4.18}$$

$$\frac{dz}{dt} = xy - \beta z \tag{4.19}$$

The modeling of the Lorenz equation at slightly changing initial conditions demonstrates that the solutions form a butterfly-shaped set of points (Figure 4.11). Years later, in his presentation "Does the flap of a butterfly's wings in Brazil set off a tornado in Texas?" in 1972, he named the phenomenon butterfly effect.

A good example of a "butterfly effect" in physiology is cancer. There is a long latency time, about 8 years, between a triggering change in a single cell and the detectable about 1–3 mm tumor. It is impossible to determine the triggering factor based on the outcome. It is also impossible to predict the outcome of a triggering factor even if it is known.

FIGURE 4.10 The states of a nonlinear system at slightly altered initial conditions.

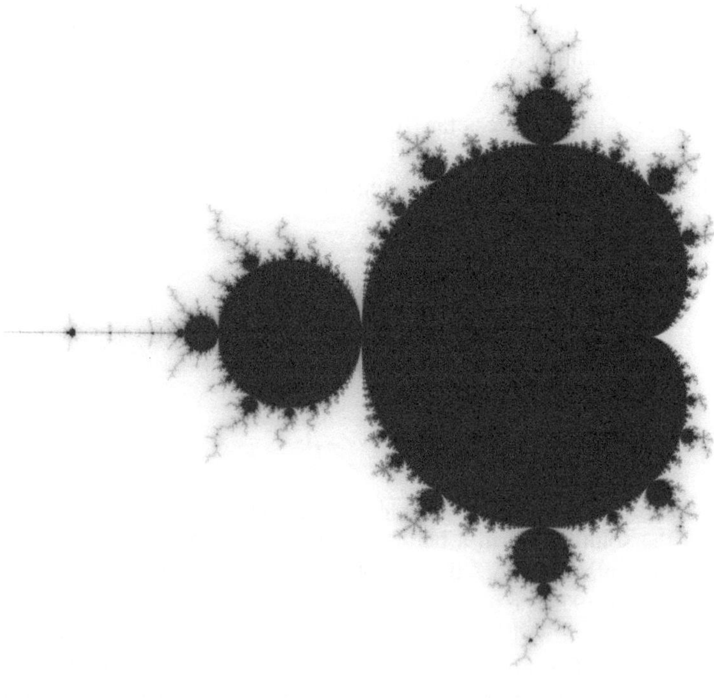

FIGURE 4.11 Mandelbrot fractals.

In the 1970s, Benoit Mandelbrot, using computer modeling of a very simple equation

$$Z_{n+1} = Z_n^{\ 2} + C \tag{4.20}$$

created a Mandelbrot set describing fractals. Fractals are a scale-invariant structure where the system replicates itself at different scales (Mandelbrot, 1977).

A maple leaf is an illustrative image of the fractality of a natural system. Fractality also appears in the structure of plant roots, leaf vessels, blood vessels, neural networks, and many others. A chaotic nature underlies many processes, including fluid flow, weather, climate, and biology.

The hypothesis that biological tissues have nonlinear properties and can produce nonlinear radiofrequency energy conversion has been tested in the experiment. The ability of living tissue to demodulate radiofrequency carriers and generate baseband in a doubly resonant cavity, possible only in the case of the nonlinear character of the transformation, was tested in a range of biological preparations (Balzano et al., 2008; Kowalchuk et al., 2010). No consistent second harmonic of the incident CW signals was detected. These results reject the nonlinearity hypothesis and support the conclusion that tissues behave as a linear medium concerning electromagnetic fields.

However, the nonlinear character of the brain response to radiofrequency radiation has become evident in the experiment by Hinrikus et al. (2011). The maximums of excitation by low-frequency modulated radiation occur not only at the half-frequency but also at the quarter-frequency of the modulation frequency, which is possible only in the case of nonlinearity of the excitation described by a nonlinear wave equation.

The discrepancy between the two results can be explained by considering the difference between the characters of the effects investigated. The production of nonlinear radiofrequency energy conversion is an electromagnetic process described by formulas derived from linear Maxwell equations. In this case, the tissue is involved only as an element with nonlinear properties, and its electric permittivity should be a nonlinear parameter depending on field strength. That is not the case. The experiment by Hinrikus et al. is aimed at exciting natural biological oscillation governed by the nonlinear Hodgkin–Huxley equation.

In this case, not the impact of the medium on the radiation but the impact of radiation on the medium has been revealed.

In general, the linear Maxwell equations describe properties of electromagnetic fields and waves in tissues. Electromagnetic radiation is known to be related to dielectric polarization in a medium linearly dependent on electric field strength (Eq. 1.10). Therefore, tissues are linear with respect to electromagnetic radiation. The nonlinear Hodgkin–Huxley equation describes the electromagnetic mechanism of neurophysiological processes. Different parameters in the Hodgkin–Huxley equation depend nonlinearly on the electric field (Eqs. 4.6–4.11). The electromagnetic field can affect the outcome of the nonlinear equation in different ways. We can conclude that the nonlinearity becomes evident in neurophysiologic processes triggered by microwave radiation but not in the properties of electromagnetic radiation in tissues.

4.7 CONCLUSIONS

The principal differences between nonlinear bioelectromagnetic processes in the brain and linear media (Chapters 1 and 2) determine the differences in the impacts of external electromagnetic fields on living and non-living systems.

- Living systems are nonlinear. The nonlinear Hodgkin–Huxley equations describe nerve pulse propagation. Biological processes like the excitation of neurons, nerve pulse propagation, bioelectric oscillations in the brain, and others are nonlinear. The response of biological systems to external electric fields is nonlinear.
- Nonlinear phenomena do not appear in a linear non-living medium and a living medium if only linear processes (dielectric polarization) are involved.
- Since the brain is a nonlinear system, the EEG signal reflecting brain oscillations is nonlinear and can be most completely described by nonlinear markers.
- The nonlinear nature of the neuronal activity makes possible the impact of even a weak electric field on the brain functioning.

REFERENCES

Adey WR, Lawrence AF, (1984). *Nonlinear electrodynamics in biological systems.* Springer, New York.

Ahmadlou M, Adeli H, Adeli A, (2012). Fractality analysis of frontal brain in major depressive disorder. *Int. J. Psychophysiol.* 85:206–211.

Aly AA, Cheema MI, Tamba Wala M, Laterza R, Zhou E, Rathnabharathi K, Barnes FS, (2008). Effects of 900-MHz radio frequencies on the chemotaxis of human neutrophils in vitro. *IEEE Trans Biomed Eng.* 55:795–797.

Bachmann M, Päeske L, Kalev K, Aarma K, Lehtmets A, Ööpik P, Lass J, Hinrikus H, (2018). Methods for classifying depression in single-channel EEG using linear and nonlinear signal analysis. *Comput Methods Programs Biomed.* 155:11–17.

Balzano Q, Hodzic V, Gammon RW, Davis CC, (2008). A doubly resonant cavity for detection of RF demodulation by living cells. *Bioelectromagnetics.* 29:81–91.

Barnes FS, Hu C-LJ, (1980). Nonlinear Interactions of electromagnetic waves with biological materials. In: PLE Uslenghi, editor. *Nonlinear electromagnetics*, pp. 391–426. Academic Press, New York.

Bartsch RP, Kang KL, Liu AB, Plamen CI, (2015). Network Physiology: How Organ Systems Dynamically Interact. *PLoS One.* 10(11):e0142143.

Bassett DS, Meyer-Lindenberg A, Achard S, Duke T, Bullmore E, (2006). Adaptive reconfiguration of fractal small-world human brain functional networks. *Proc Natl Acad Sci U S A.* 103(51):19518–19523.

Berger H, Gray CM, (1929). Uber das Elektroenkephalogramm des Menschen. *Arch Psychiat Nervenkr.* 87: 527–570.

Buchholtz F, Schinor N, Schneider FW, (2002). Stochastic nonlinear dynamics: how many ion channels are in a single neuron? *J Phys Chem B.* 106(19). doi: 10.1021/jp0120662.

Buzsáki G, Draguhn A, (2004). Neuronal oscillations in cortical networks. *Science.* 304: 1926–1929.

Buzsáki, G, (2006). *Rhythms of the brain.* Oxford University Press, Oxford.

Croft RJ, Hamblin DL, Spong J, Wood AW, McKenzie RJ, Stough C, (2008). The effect of mobile phone electromagnetic fields on the alpha rhythm of human electroencephalogram. *Bioelectromagnetics.* 29(1):1–10.

Croft RJ, Leung S, McKenzie RJ, Loughran SP, Iskra S, Hamblin DL, Cooper NR, (2010). Effects of 2G and 3G mobile phones on human alpha rhythms: resting EEG in adolescents, young adults, and the elderly. *Bioelectromagnetics.* 31(6):434–444.

Curcio G, Mazzucchi E, Della Marca G, Vollono C, Rossini PM, (2015). Electromagnetic fields and EEG spiking rate in patients with focal epilepsy. *Clin Neurophysiol.* 126(4):659–666.

Fingelkurts AA, Fingelkurts AA, Rytsälä H, Suominen K, Isometsä E, Kahkonen S, (2007). Impaired functional connectivity at EEG alpha and theta frequency bands in major depression. *Hum. Brain Mapp.* 28(3):247–261.

Gerstner W, Kistler WM, Naud R, Paninski L, (2014). *Neuronal dynamics from single neurons to networks and models of cognition.* Cambridge University Press, Cambridge.

Hinrikus H, Bachmann M, Lass J, Tomson R, Tuulik V, (2008a). Effect of 7, 14 and 21 Hz modulated 450 MHz microwave radiation on human electroencephalographic rhythms. *Int J Radiat Biol.* 84(1):69–79.

Hinrikus H, Bachmann M, Lass J, Karai D, Tuulik V, (2008b). Effect of low frequency modulated microwave exposure on human EEG: Individual sensitivity. *Bioelectromagnetics.* 29(7):527–538.

Hinrikus H, Bachmann M, Karai D, Klonowski W, Lass J, Stepien P, Stepien R, Tuulik V, (2011). Higuchi's fractal dimension for analysis of the effect of external periodic stressor on electrical oscillations in the brain. *Med Biol Eng Comput.* 49(5):585–591.

Hinrikus H, Lass J, Karai D, Pilt K, Bachmann M, (2015). Microwave effect on diffusion: a possible mechanism for non-thermal effect. *Electromagn Biol Med.* 34(4):327–333.

Hodgkin AL, Huxley AF, (1952). A quantitative description of membrane current and its application to conduction and excitation in nerve. *J Physiol.* 117(4):500–44.

Hosseinifard B, Moradi MH, Rostami R, (2013). Classifying depression patients and normal subjects using machine learning techniques and nonlinear features from EEG signal. *Comput Methods Programs Biomed.* 109:339–345.

Hutcheon B, Yarom Y, (2000). Resonance, oscillation and the intrinsic frequency preferences of neurons. *Trends Neuros.* 23:216–222.

Izhikevich EM (2007). *Dynamical systems in neuroscience.* MIT, Cambridge, MA.

Kaplan D, Glass L (1995). *Understanding nonlinear dynamics.* Springer-Verlag, New York.

Kowalczuk C, Yarwood G, Blackwell R, Priestner M, Sienkiewicz Z, Bouffler S, Ahmed I, Abd-Alhameed R, Excell P, Hodzic V, Davis C, Gammon R, Balzano Q, (2010). Absence of nonlinear responses in cells and tissues exposed to RF energy at mobile phone frequencies using a doubly resonant cavity. *Bioelectromagnetics.* 31:556–565.

Lorenz E, (1963). Deterministic nonperiodic flow. *J. Atmos. Sci.* 20:130–141.

Lorenz EN, (1972). Predictability: does the flap of butterfly's wings in Brazil set off a tornado in Texas? by Presented before the American Association for the Advancement of Science, December 29, 1972.

Lin A, Liu KK, Bartsch RP, Ivanov PC (2016). Delay-correlation landscape reveals characteristic time delays of brain rhythms and heart interactions. *Phil. Trans. R. Soc. A* 374:20150182.

Lustenberger C, Murbach M, Tushaus L, Wehrle F, Kuster N, Achermann P, Huber R, (2015). Inter-individual and intra-individual variation of the effects of pulsed RF EMF exposure on the human sleep EEG. *Bioelectromagnetics.* 36(3):169–177.

Malmivuo J, Plonsey R, (1995). *Bioelectromagnetism.* Oxford University Press, New York.

Mandelbrot B, (1977). *Les objects fractals: forme, hasard et imension.* Flammarion, Paris.

Muler AL, Markin VS, (1977). Electrical properties of anisotropic neuromuscular syncytia. II. Distribution of a flat front of excitation. *Biofizika.* 22(3):518–522.

Nunez PL, (1981). *Electric fields of the brain: the neurophysics of EEG.* Oxford University Press, New York.

Päeske L, Hinrikus H, Lass J, Raik J, Bachmann M, (2020). Negative correlation between functional connectivity and small-worldness in the alpha frequency band of a healthy brain. *Front Physiol.* 11:910. doi: 10.3389/fphys.2020.00910.

Päeske L, Uudeberg T, Hinrikus H, Lass J, Bachmann M, (2023). Correlation between electroencephalographic markers in the healthy brain. *Sci Rep.* 13(1):6307. doi: 10.1038/s41598-023-33364-z.

Rowbottom M, Susskind C, (1984). *Electricity and medicine. History of their interaction*, San Francisco Press, San Francisco.

Schnitzler A, Gross J, (2005). Normal and pathological oscillatory communication in the brain. *Nat Rev Neurosci.* 6:285–296.

Steriade M, Timofeev I, (2003). Neuronal plasticity in thalamocortical networks during sleep and waking oscillations. *Neuron.* 37:563–576.

Strogatz S, (2015). *Nonlinear dynamics and chaos: with applications to physics, biology, chemistry, and engineering.* Westview Press, Boulder CO.

Thompson JMT, Stewart HB, (2002). *Nonlinear dynamics and chaos.* John Wiley & Sons, New York.

Vecchio F, Babiloni C, Ferreri F, Buffo P, Cibelli G, Curcio G, van Dijkman S, Melgari JM, Giambattistelli F, Rossini PM, (2010). Mobile phone emission modulates interhemispheric functional coupling of EEG alpha rhythms in elderly compared to young subjects. *Clin Neurophysiol.* 121(2): 163–171.

Part II

Understanding the mechanism of low-level microwave radiation effect

From the principles of electromagnetism to experimental results

Imagination is more important than knowledge.

Albert Einstein

DOI: 10.1201/9781003729976-6

5 The physical mechanisms of low-level microwave radiation effects

HIGHLIGHTS

- Coherence of the radiation is crucial in non-thermal mechanisms of low-level radiation.
- The cumulative effect of small disturbances enables the coherent microwave radiation to overcome the thermal noise limit *kT*.
- Microwave radiation can affect the majority of physiological processes due to its impact on diffusion.
- It is possible to excite brain bioelectrical oscillations at predetermined frequencies using low-level microwave radiation.
- The effect of radiofrequency radiation on the brain would not be detectable without low-frequency modulation of the radiation. The modulation is crucial in revealing the effect.
- The non-thermal mechanism has no physical threshold.

5.1 INTRODUCTION

The non-thermal effects of microwave radiation in biological systems have been debatable for decades. Electromagnetic radiation is a physical factor. The primary physical limitations of the concept of the nonthermal mechanism of microwave radiation are the following:

- The energy of the quanta of the radiation at radiofrequencies is not enough to remove electrons from the atom and ionize atoms or molecules. The energy of electromagnetic radiation *hf*, where *h* is the Planck constant, becomes equal to the thermal energy *kT*, where *k* is the Boltzmann constant, in the infrared region at a frequency of about 7000 GHz.
- The radiofrequency electric field inside living tissue is significantly less than intermolecular electric fields, approximately 10^6 V/m. Therefore, the direct impact of microwave radiation on body structures at the atomic and molecular levels is unlikely.
- The intramolecular electric fields, despite being some orders of magnitude weaker than intermolecular fields, are still much stronger than the fields from the technical sources of radiofrequency radiation in the human body.

DOI: 10.1201/9781003729976-7

- The energy of radiofrequency radiation is much less compared to the level of thermal energy at the temperature of $T = 300$ K. The thermal noise limit has made it complicated to explain nonthermal effects of low-level radiofrequency radiation.

The only possible effect at radiofrequencies is magnetic. The spin of electrons bound to molecules is the only mechanism not significantly affected by temperature, and the precession of spins by external magnetic fields is not limited by thermal noise. Therefore, the spin orientation in the recombination of radical pairs may constitute a magnetic mechanism for the weak low-frequency and radiofrequency fields (Adair, 1999; Barnes and Greenebaum, 2015). The sensitivity of living organisms to magnetic fields is supported by several studies (Barnes and Greenebaum, 2018, 2020; Henshaw and Philips, 2025).

Additional limitations are related to the physics of microwaves. Microwave radiation cannot cause any regular changes in the movement of cells and ions due to their small cross-section of absorption (Eq. 2.35) because the wavelength of the radiofrequency wave is much larger than the dimensions of a cell or ion. The resonances in biological structures are damped by inertial properties and viscosity of the liquid medium and are unlikely (Adair, 2002). The microwave frequencies are much higher than physiological frequencies and cannot directly affect physiological processes.

The main reasons listed above have made the explanation of the mechanisms of microwave effects in humans so complicated over the decades. Due to frequency limitations, the magnetic mechanism is unlikely in microwaves. Therefore, for over a century, only the thermal mechanism of the radiofrequency radiation effect has been taken into account and commonly accepted (Barnes, 1992; Foster, 2000; Lawrence et al., 2003; Polk and Postow, 1986). To date, there is no commonly accepted non-thermal mechanism of the microwave effect.

However, several experimental findings cannot be explained by the thermal mechanism of microwave radiation. The first studies by Bawin et al. (1973) and Adey et al. (1982) demonstrated that low-frequency modulation of microwave radiation is necessary to reveal the effects on the central nervous system and calcium efflux. The effective level of energy of the radiation decreases with modulation; therefore, modulation is expected to reduce the thermal effect. The impact of modulation, increasing the effect of microwave radiation, cannot be explained by the thermal effect.

During the following decades, numerous authors have reported the effects of low-level modulated microwave radiation, and non-thermal mechanisms have been proposed by various scientists. Whereas the energy of weak electromagnetic fields is much less than thermal energy, the nonthermal effects are expected to be veiled by thermal noise and are not likely. Despite that, experimental studies performed at low-level microwave radiation have reported positive results.

This chapter proposes a novel approach and demonstrates that not a special mechanism but dielectric polarization, a well-known and commonly accepted phenomenon, can trigger small initial changes sufficient to result in significant alterations in nonlinear living systems and cause non-thermal effects in microwaves.

5.2 NONTHERMAL MECHANISMS OF MICROWAVE RADIATION IN BIOLOGICAL SYSTEMS: POSSIBLE MODELS AND LIMITATIONS

To resolve the discrepancy between experimental results and the thermal mechanism, several attempts have been made to find an adequate non-thermal mechanism. Over the decades, very different mechanisms have been proposed by many scientists to explain the low-level microwave radiation effect. The reviews by Adair (2003), Challis (2005), Sheppard et al. (2008), and Apollonio et al. (2013) aim to evaluate the possibility and limitations of non-thermal effects of low-level radiofrequency radiation.

The fundamental analysis by Robert Adair (2003) aims to evaluate the direct physical mechanism of electromagnetic fields on biological and physiological structures. The analysis is based on the condition that the final physiological consequences of the primary interaction must vary linearly with the energy density $w = \varepsilon E^2$. The criterion that the power density w_i of the radiation causing a direct physical effect must be greater than that from the thermal noise in a biological system, the ratio $w_i/kT>1$, has been used as the criterion of the possibility of the effect. The *S/N* of a direct field effect estimated on the model of a spherical proton moving in a viscous fluid is of the order of magnitude 10^{-13} at the level of electric field strength $E = 200$ V/m. The criterion has been used to evaluate various direct interactions, including rotational motion, linear motion, opening voltage-gated membrane channels, electrostrictive forces, dipole–dipole interaction, resonances, and radical pairs recombination. Possible direct electromagnetic interactions are unlikely at low-level microwave radiation in all discussed mechanisms where biology is expected to be affected by the motion of charged elements generated by the electric field. The general conclusion states that radiofrequency radiation with intensity less than 10 mW/cm^2 (electric fields less than 200 V/m) has proved unlikely to significantly affect physiology through nonthermal mechanisms. Identical criteria have been used in other reviews (Challis, 2005; Sheppard et al., 2008).

The results of quantitative analysis of fundamental physical mechanisms of radiofrequency radiation effects in biological molecules and processes have been presented in a comprehensive review by Sheppard et al. (2008). The review focuses on the physical principles that guide quantitative assessment of mechanisms applicable for exposures at or below the level of endogenous electric fields, generally from 1 to 10^2 V/m, on conditions where temperature increases are insignificant, less 1 K. Sensitivity and time constants of various possible mechanisms including signal averaging, coherent detection, nonlinear dynamical systems, radical pairs, role of magnetite, and many others have been discussed. The quantitative assessment has led to the conclusion that the proposed mechanisms can be acceptable at lower frequencies but become unrealistic at microwave frequencies because of serious physical restrictions in the time constant (Sheppard et al., 2008). None of these special mechanisms has been experimentally validated in the microwave frequency band.

It has been concluded that the frequency limitations also make unlikely other discussed biological effects such as changes in **protein conformation** (Astumian, 2003; Bohr and Bohr, 2000; Laurence et al., 2000, 2003) and **changes in binding ligands** like Ca2þ to cell receptor proteins (Chiabrera et al., 2000). The effects are plausible at lower frequencies but not at microwaves (Challis, 2005). The analysis by Adair (2003) concludes that none of a set of interactions of weak fields can affect biology on the **molecular scale** and that such weak fields are quite unlikely to generate significant effects in their interactions with larger biological elements such as cells.

The theoretical model for non-thermal mechanisms, **resonance-type interactions** between microwave waves and biological systems, such as microtubules, has been proposed by Herbert Frölich (1968). However, this theoretical model has not been supported by experiments. The resonances in microtubes or other biological structures are overdamped due to the viscosity of biological systems (Adair, 2002; Sheppard et al., 2008) and are not likely.

The **effect on cell membrane potential** is possible because the electric potential on the membrane is much larger than the average field. Any small alteration in the level of the electric potential of the average field caused by an external field results in voltage drops across the membrane and might be large enough to give rise to nonlinear effects including rectification. However, the capacity and related time constant of a membrane limit the frequency of the effect to about 10 MHz. The mechanism is unlikely at microwave frequencies (Challis, 2005; Sheppard et al., 2008).

Radiofrequency electric field produces **oscillating dipole moments** within cells, which enhance the attraction between cells and cause the pearl-chain effect (Schwan, 1985). However, the calculations show that an electric field strength of about 125 V/m is required to produce the effect at frequencies less than 100 MHz, and at microwave frequencies, the effect is unlikely (Adair, 2002; Challis, 2005; Sheppard et al., 2008).

The magnetic field effect can be related to the **radical pairs** mechanism. A radical is a part of a molecule with an unpaired valence electron or hole. Radicals are reactive chemically. The reactive structures may join to reform the molecule, eliminating reactivity, but only in a condition of a singlet state of the unpaired electrons attached to the two radicals. The external magnetic field can affect the state of a radical. The hyperfine splitting of spins in most biomolecules is less than 100 MHz. The randomization of the orientations of the electron spins (spin relaxation) is expected to be too fast for frequencies higher than 100 MHz fields to have a significant effect on the state of the radical pairs. It seems unlikely that radiation at microwave frequencies could lead to a significant increase in the concentration of free radicals (Adair, 2003; Challis, 2005; Sheppard et al., 2008).

Nonlinear transformation and demodulation of microwave energy by living cells have not been supported by the experiment (Balzano et al., 2008; Kowalczuk et al., 2010). In this study, the nonlinear radiofrequency energy conversion in a range of biological preparations has been evaluated, testing the hypothesis that living tissue can demodulate microwave carriers and generate baseband signals. No consistent second harmonic of the incident CW signals was detected. Therefore, these results do not support the hypothesis that living cells can demodulate radiofrequency energy since second harmonic generation is the necessary and sufficient condition for demodulation (Kowalczuk et al., 2010).

The negative result may be related to the fact that the demodulation and generation as electromagnetic processes are described by linear electromagnetic field equations not involving biology (Section 4.5). In this case, biological tissues behave as a linear medium regarding electromagnetic fields due to the linearity of the relationship between fields and tissues (Eqs. 1.5, 1.7, 1.9). The nonlinear response can be evident in living biological processes described by nonlinear, e.g., Hodgkin–Huxley, equations. The results of the experiment demonstrate that tissues are linear as a medium, and the nonlinearity appears only when biological processes are involved, e.g., at the excitation of biological oscillations (Hinrikus et al., 2011).

Apollonio et al. (2013) present a critical literature review of the models of the interaction mechanisms, together with an overview of all the publications finding positive results for in vitro and in vivo studies. The authors of the review have used a multiscale methodology, discussing effects from atoms to organs, to perform a comprehensive study of the nonthermal effects. The authors have pointed out five plausible hypotheses: (1) selective microthermal heating; (2) nonthermal effects on reactions; (3) action on hydrogen bonds in the hydration layer; (4) radical pair mechanism; and (5) iron-ion mediated mechanisms. However, such a selection is not limited to physical effects but includes some biochemical–biological effects like (1) protein synthesis and (2) oxidative stress. It has been concluded that selective microthermal heating can, in principle, justify all the effects reported. The other mechanism can only partially explain some of the effects reviewed. However, microthermal heating lacks experimental support. The selected mechanisms are not always the primary effect of radiation. Nonthermal effect on reactions can be explained as a consequence of diffusion-increased microwave radiation (Section 5.4). As concluded by the earlier reviews, magnetic ion–ion-medicated radical pair mechanisms are plausible at lower frequencies but not at microwave frequencies higher than 100 MHz (Adair, 2003; Challis, 2005; Sheppard et al., 2008).

Based on the thermodynamic criterion of the plausibility of the effect, which requires a higher level of radiation intensity compared to thermal energy, all possible mechanisms proposed by various scientists have been ruled out by the above reviews. It is important to pay attention that this criterion is a result of analysis performed in the condition of a linear relationship between the effect and the radiation field power density proportional to E^2. The criterion, based on a comparison with the level of thermal fluctuations, makes the derived conclusions expectable.

Another approach based on the linear relationship between the effect and electric field strength has been proposed by Hinrikus et al. (2018). The proposed model is based on the presumption that the origin of the effect is the dipolar polarization of tissues by electromagnetic radiation. The polarization is a linear function of the electric field strength (Eq. 1.10). The plausibility of the model is supported by the fact that dipolar polarization exists at low-level exposure and has no principal threshold.

The polarization is expected to produce small alterations in the position of charges inside tissues. The effect can become remarkable and measurable due to the cumulation of simultaneous in-phase motions over a volume of half-wavelength size. These minimal alterations do not cause a biological effect but change a little the initial and boundary conditions for the nonlinear dynamic equations describing the physiological processes (Section 4.5). The small alterations in initial and boundary conditions can result in remarkable changes in the solution of nonlinear dynamic equations (Section 4.5). Therefore, the energy that produces the radiofrequency radiation effect is coming not from the radiation but from the metabolism supporting physiological processes. Therefore, the nonthermal mechanism of radiofrequency radiation becomes possible.

The nonthermal mechanism is not an obligatory related to "low-level" effect; it also occurs at high levels of radiation. Two different microwave effects, thermal and nonthermal, cooccur parallel. The difference between the effects of thermal and microwave heating is well-known in chemistry and technology (Wang et al., 2023; Zong et al., 2003).

5.3 THERMAL EFFECT

The power of radiofrequency radiation absorbed in the human body can be transferred to thermal power (Eq. 1.29). The increase of thermal energy over time interval Δt results in the increase of temperature ΔT

$$\sigma E^2 \Delta t = k \Delta T \tag{5.1}$$

where $k = 1.38 \times 10^{-23}$ [J/K] is the Boltzmann constant.

The increased temperature enhances the kinetic energy of particles inside tissues. The increase in temperature ΔT results in the increase of kinetic energy of the particles with the mass of m in three-dimensional space as

$$\frac{3}{2} k \Delta T = \frac{mv^2}{2} \tag{5.2}$$

where v is the velocity of a particle.

The thermal effect is caused by the absorbed in the tissues electromagnetic energy. The absorbed electric energy causes a random Brownian motion. It is not related to the orientation of electric forces and therefore not sensitive to the phase of the electric field.

The increase in temperature, stimulating the motion of particles, affects positively many physiological processes in tissues. Increasing temperature accelerates biochemical reactions, enhances diffusion, accelerates metabolism, supports the removal of harmful substances, etc. The effect of radiofrequency heating, the so-called d'Arsonval effect, was discovered even earlier than the application of radiofrequency for communication. Radiofrequency radiation is successfully used in medical therapy.

On the contrary, too high a temperature can lead to damage to tissues. Human physiology has mechanisms to regulate the temperature inside a body and can stabilize the rise of temperature within some temperature interval. The rise in temperature behind the limits of temperature stabilization is potentially harmful and should be avoided. The Arrhenius equation, published by Svante Arrhenius in 1899, describes the rate of thermal reaction K as an exponential function of the ratio of molar activation energy E_a to thermal energy kT

$$K = Ae^{-\frac{E_a}{kT}} \tag{5.3}$$

where A is the Arrhenius factor.

The Arrhenius model describes exponential dependence on temperature related to the radiation power density.

Based on the Arrhenius model, John Pearce has described thermal damage to tissue, burns, and cell death with temperature that occurs due to the thermal denaturation of proteins (Pearce, 2018). Thresholds for thermal damage are functions of both the temperature and exposure time. The Arrhenius model establishes no theoretical threshold of thermal effect. Any increase in temperature increases the rate of thermal reaction. Human body mechanisms can repair tissue damage and restore normal functioning of tissues at small temperature increases. The current health protection regulations are aimed at limiting the rise of temperature to 1°C, which is considered not harmful.

The increasing temperature accelerates chemical reactions, and heating is widely used in chemical technology. The advances in microwave chemistry during the last decades have shown that heating by radiofrequency radiation is more effective than traditional thermal heating. The various applications of microwave heating reported by Galema (1997), Zong et al. (2003), Hoz et al. (2005), Antonio and Deam (2007), and Huang et al. (2025) have supported the possibility of reducing several times the temperature and heating time and decreasing the energy consumption in chemical reactions and materials processing (Wang et al., 2023; Zong et al., 2003).

5.4 COHERENCE OF RADIATION AND CUMULATIVE EFFECT

The dielectric polarization of tissues (Section 1.2) determines the dielectric permittivity of the tissues (Section 3.1) and, through that, properties of electromagnetic fields in tissues. The presence of a radiofrequency electric field causes the rotational movement of dipolar molecules and produces orientational polarization that dominates at radiofrequencies.

The energy of intermolecular bonding between tissue molecules is less than the intermolecular energy but still much higher than the energy of radiofrequency radiation. However, the rotation of molecules related to radiofrequency radiation is known to produce restructuring or even breaking bonding between the molecules.

A principal question of how a relatively weak radiofrequency field can affect charges in a medium and rotate dipolar molecules with bonding energies higher than the energy of the radiation was answered shortly by Zahn in his book (Zahn, 2003). Due to the large number of atoms and molecules in a medium, the synchronization of field-induced small imbalances in atoms and molecules leads to a strong cumulative impact and high effectiveness in summarizing the imbalances. The cumulative effect is known and performs important tasks in chemistry and biology (Silberberg, 2012). Relatively weak hydrogen bonds are known to keep together biomolecules due to the cumulative effect.

In the case of dielectric polarization, the electric field affects a huge number of particles simultaneously. In this case, the cumulative summarizing should be considered in estimating the energy of the electric field effect. Let us discuss a cumulative effect using a water model for tissues. The number of water molecules in one gram is determined by the Avogadro number 6.022×10^{23} as $6.022 \cdot 10^{23}/18 = 3.455 \cdot \times 10^{22}$

where 18 is the molecular weight of water. The number of dipolar water molecules performing rotational movement in a volume of a half-wave 1 mm^3 is of the order of magnitude 10^{22}. The number of synchronously affected molecules is limited to a half-wave of the radiation. At lower frequencies, this number and the related cumulative effect increase. This may be one of the reasons why the electric permittivity increases with decreasing frequency (Section 3.2).

The energy of thermal noise kT is much higher than the energy of microwave radiation. The criterion of plausibility of nonthermal effect estimated by Adair (2003), signal-to-noise ratio $S/N_p \approx 3 \times 10^{-13}$, is based on a model describing the rotational movement of a single spherical particle in the electric field at $\boldsymbol{E} = 200$V/m (10 mW/cm^2). The summarizing of n synchronous coherent small alterations is proportional to n. The cumulative effect of summarizing random alterations is proportional to $n^{1/2}$. The cumulative effect of summarizing changes in the criterion proposed by Adair for evaluating the possibility of microwave effects leads to $S/N \approx 3 \times 10^{-2}$ in the case of noncoherent thermal radiation and to $S/N \approx 3 \times 10^9$ in the case of coherent radiation. This estimate shows that coherent radiation is much more likely to produce an effect than noncoherent radiation. These numbers show that the cumulative effect of coherent microwave radiation can overcome the thermal limit cwt. It is not a surprising conclusion but presents some quantitative support for the well-known phenomenon. The ability of low-level electromagnetic radiation to produce dipolar polarization and various biological effects is supported by many experimental results.

The effectiveness of the cumulative impact depends on the stability of the phase of the electromagnetic wave. At microwaves, the wavelengths are larger than the dimensions of molecular or cellular structures in tissues. The initial character of the phase distribution in the time domain is determined by the source of radiation. Due to the determined value of electromagnetic field phase velocity, the temporal phase distribution is transferred to the spatial distribution of the phase that determines the instantaneous directions of the electric forces at different points. The coherence of the radiation is very important to ensure the co-direction and synchronicity of forces and related small changes (Figure 5.1).

In the case of coherent radiation (Figure 5.1B), the phase of the electric field is constant, and electric forces are oriented in one direction. Within a wavelength, all charges in a medium are affected by synchronous forces (Figure 5.1B). The vectors of electric forces $\boldsymbol{F}_i$ are parallel to each other and the sum of n vectors:

$$\boldsymbol{F} = \boldsymbol{F}_1 + \boldsymbol{F}_2 + \boldsymbol{F}_3 + \equiv\ + \boldsymbol{F}_n. \tag{5.3}$$

In the case of equal values of the forces, the sum is proportional to n.

In the case of noncoherent radiation, the phase is a random function of time. The random phase determines the random character of electric forces distributed in random directions (Figure 1A). In the case of large n, the pairs of orthogonal vectors always exist and the sum of n vectors:

$$\boldsymbol{F} = \sqrt{\boldsymbol{F}_1^2 + \boldsymbol{F}_2^2 + \boldsymbol{F}_3^2 + \ldots + \boldsymbol{F}_n^2} \tag{5.4}$$

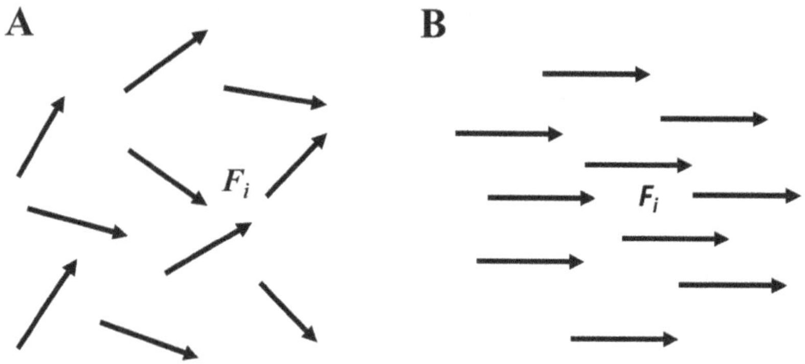

FIGURE 5.1 Distribution of electric forces inside a homogeneous tissue. (A) Incoherent radiation; (B) coherent radiation.

In the case of equal lengths of initial vectors, the sum is proportional to $n^{1/2}$. In the general case of randomly distributed nonequal n vectors, the dependence on n keeps its character despite more complicated summarizing. For example, Dill and Bromberg showed in 2002 that the expected translation distance after n steps of random motion should be of the order of $n^{1/2}$.

The coherence of electromagnetic radiation is crucial for causing a cumulative effect sufficient to produce dielectric polarization of living tissues and a nonthermal effect, as shown in Section 5.1. The cumulative effect of noncoherent radiation is insufficient to support dipolar polarization and cannot cause a nonthermal effect.

The synchronous cumulative impact of coherent radiofrequency radiation in tissues has been convincingly confirmed by the phenomenon of dielectric polarization: a very weak electric field, compared to intra- or intermolecular fields, can create a measurable electric polarization of the tissues (Section 3.1).

The thermal and non-thermal effects of radiofrequency radiation can occur together despite their different nature:

- Non-thermal effect is generated by coherent radiation, causing simultaneous regular movement of particles, and it can take place at a constant temperature. With increasing temperature, random thermal motion rather decreases the effect, disturbing the regular motion.
- The thermal effect is produced by increased temperature, which causes random motion of particles. In parallel, due to the coherent nature of radiation, simultaneous regular alterations take place (non-thermal effect). The thermal effect caused by coherent radiation differs from traditional thermal heating and is stronger.

The presence of both mechanisms in thermal effect has been most clearly demonstrated in microwave chemistry and material technology, where the heating by radiofrequency radiation has been demonstrated to be much more effective than traditional thermal heating.

5.5 EFFECT ON HYDROGEN BONDS

Attractive forces between positive and negative charges of the molecules cause hydrogen bonds between dipolar molecules (Gilli and Gilli, 2009). The energies associated with hydrogen bonds range from a few to tens of *k*its Hydrogen bonds have about a tenth of the strength of an average covalent bond and are constantly broken and reformed in liquid water. They determine the viscosity and boiling temperature of water. Hydrogen bonds contribute to the properties of proteins and other biomolecules. Hydrogen bonds are responsible for the overall stability of the DNA double helix structure. Although intermolecular attractive forces of hydrogen bonding are rather weak, the consolidated power of the huge amount of hydrogen bonds in DNA provides the stability of the molecule.

Hydrogen bonds are produced by the electric forces between charges. Therefore, the bonds between molecules are affected by the external electric field (Section 1.1). Despite the energy of radiofrequency radiation being a much lower bonding energy, coherent radiation can cause, due to the cumulative effect, the rotational movement of dipolar molecules in a medium. The rotation of dipolar molecules can disturb the structure of hydrogen bonding (Figure 5.2). The reorientation of molecules by radiofrequency radiation results in the restructuring or even breaking of the bonds between the molecules.

In a non-polarized medium (Figure 5.2A), the structure of attractive electric forces between water molecules is determined by the mutual orientation of molecules: each positively charged hydrogen ion is attracted by the nearest negatively charged oxygen ions in another molecule, and the attraction between positive and negative charges is maximal while the charges are oriented to each other. Hydrogen bonds are distributed in all directions. The random Brownian motion constantly breaks and reforms the hydrogen bonds in a liquid medium.

The external electric field, causing electric polarization of the medium, aligns the molecules with the field direction and reorganizes the structure of hydrogen bonds (Figure 5.2B). The rotation of dipolar molecules by a radiofrequency electric field results in the realignment of dipolar molecules. The mutual optimal orientation between positive and negative charges is disturbed. Even partial realignment of molecules by the field causes hydrogen bonding to become bent or broken.

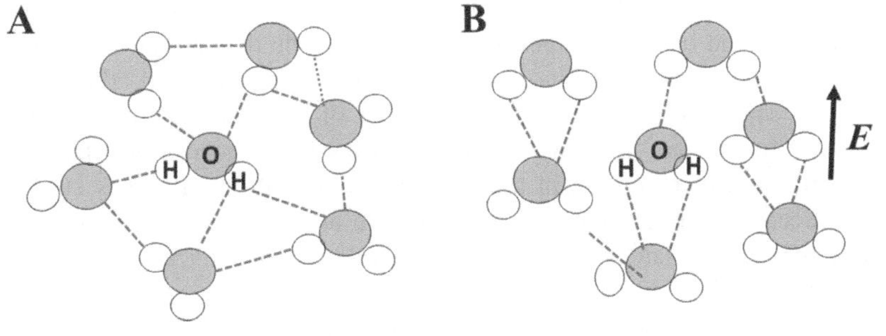

FIGURE 5.2 The structure of hydrogen bonds in water. (A) Without an external electric field; (B) in the presence of an external electric field.

The inertial properties of the hydrogen bonds are critical for the evaluation of the impact of microwave radiation. The computer simulation of hydrogen bond kinetics in liquid water performed by Luzar and Chandler (1996) has shown that the time constant of the processes is about 1 ps. The experiments by Bakker et al. (2000) have demonstrated hydrogen bond stretching dynamics in liquid water and reported a reorientation time constant of about 1 ps. The spectrum of the water hydrogen bond network has been investigated by Shi raga et al. (2017) in the wide frequency band from 0.5 GHz to 12 THz, and no frequency restrictions have been reported for the microwave radiation effect. There are no frequency limitations foreseen for the impact of microwave radiation on hydrogen bonds.

The restructuring of hydrogen bonds caused by radiation means the weakening of intermolecular bonds in cells, leading to increased levels of reactive oxygen species (ROS). The limited ability of cells to neutralize the increased levels of ROS results in oxidative stress. The relationship between low-level radiofrequency radiation, ROS, and oxidative stress has been reported in many animal and cellular studies discussed in the reviews by Yakymenko et al. (2016), Schuermann and Mevissen (2021), and Meyer et al. (2024). Oxidative stress can damage cellular components like lipids, proteins, and DNA.

The ability of microwave radiation to induce alterations in intermolecular bonding has been most clearly demonstrated in microwave chemistry. The advantages of microwave heating-supported reactions have demonstrated significant advantages compared to traditional thermodynamic heating in chemistry; microwave-assisted synthesis of novel organic, organometallic, and inorganic molecules or complexes already reported by Galema in 1997; and many other studies discussed in the recent review by Wang et al. (2023). Important advantages of microwave technology over conventional synthesis supported by traditional heating include a highly accelerated rate of the reaction and a reduction in reaction time with an improvement in the yield and quality of the product.

5.6 EFFECT ON DIFFUSION

Radiofrequency radiation causes the restructuring of hydrogen bonds with the frequency of radiation. Such "shaking" of hydrogen bonds determining the water viscosity is expected to increase diffusion.

Diffusion is responsible for many processes in humans, beginning from molecular movement and ligand–receptor interactions outside the cell. Diffusion contributes to the mechanism of the movement of oxygen, nutrients, and other molecules across the capillary walls and the movement of molecules across membranes. The neurotransmitters move across the synapses by a process of diffusion. The lateral diffusion of receptors in the plane of the membrane is a key step in regulating receptor stabilization and accumulation at synapses. Recent studies demonstrate that inflammatory processes are related to diffusion (Aronica et al., 2022; El Hajj et al., 2022).

According to the first Fick's law, the diffusion current density

$$j_D = -D\nabla c \tag{5.5}$$

where ∇c is the ion concentration gradient and D is the diffusion coefficient. Stokes–Einstein's equation gives the diffusion coefficient

$$D = \frac{kT}{\chi} \tag{5.6}$$

where χ is the friction coefficient determined by viscosity.

The diffusion coefficient increases linearly with temperature and decreases with friction related to hydrogen bonding. Diffusion is determined by the energy of random motion kT and therefore directly depends on temperature. Friction is related to hydrogen bonds and decreases with them as temperature increases. Increasing the numerator and decreasing the denominator in the fraction (Eq. 5.6) results in a very strong dependence of the diffusion coefficient on temperature.

The high-level radiofrequency radiation causing thermal effect increases diffusion. The low-level radiofrequency radiation affecting hydrogen bonding without an increase in temperature decreases viscosity and is expected to increase diffusion despite the expected effect being less than the thermal effect.

The ability of radiofrequency radiation to increase cell velocity and diffusion behind the thermal effect has been supported by an experiment by Aly et al. (2008) (Figure 5.3). The exposure to 900 MHz microwave radiation at about 0.4 V/m increased the velocity of molecules in human blood above the value observed at the same temperature by approximately 50%. The calculated temperature change resulting from the radiofrequency exposure was less than one microdegree.

The increasing velocity of molecules is related to the decreased viscosity of the blood and increased diffusion. The low-level microwave radiation affects similarly

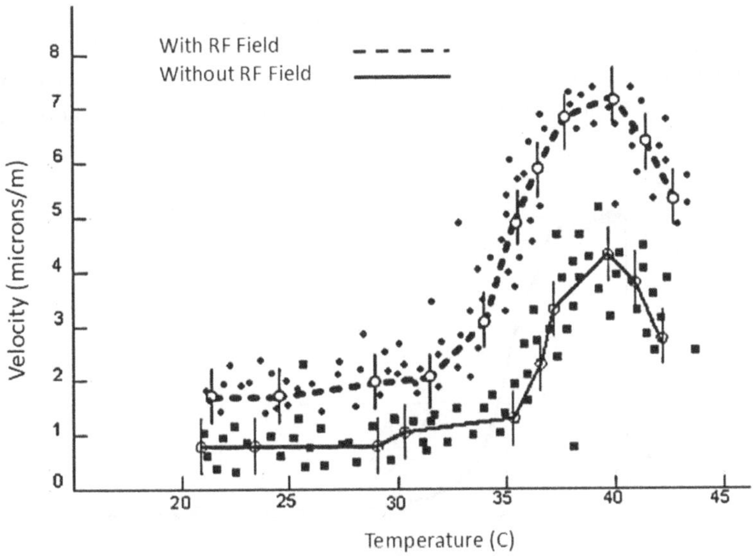

FIGURE 5.3 Cell velocity as a function of temperature with and without exposure to RF fields at 900 MHz and approximately 0.4 V/m (from Ali et al., 2008).

to the increase in temperature but at a constant temperature. The nonthermal effect can be considered a quasi-thermal effect (Hinrikus et al., 2005, Hinrikus et al., 2017).

An experiment on the investigation of the impact of radiofrequency radiation on diffusion in water at constant temperature was performed by Hinrikus et al. (2015). The experimental study was performed on the setup of two identical reservoirs filled with pure water and 0.9% NaCl solution and connected by a thin tube. Alterations of NaCl concentration in the reservoir initially filled with pure water were automatically measured using the resistance of the solution as an indicator. The 450 MHz continuous-wave radiation was applied at the calculated maximal absorption rate in the connecting tube of 0.4 W/kg. The recorded data (Figure 5.4) show a significant acceleration of the alterations in concentration caused by the radiofrequency radiation exposure at constant temperature.

The results reported in two experiments performed by independent groups of researchers and using different methods support the impact of low-level radiofrequency radiation on diffusion at constant temperature.

Diffusion affects the majority of physiological processes, beginning from molecules and cells to systems level. The response of human physiology to the alterations of diffusion is not singularly determined. So many different possible variants of the microwave radiation effect in diffusion can, in turn, cause very different changes in biology, which are difficult to relate to their primary cause, microwave radiation. Therefore, the experiments cannot directly support the impact of diffusion but only the secondary effects of the radiation on various processes related to diffusion.

The ability of a low-level coherent electromagnetic field to accelerate diffusion at a constant temperature when heating is not evident can be useful in the interpretation of experimental results. The effect of the increase in calcium ion efflux through cell membranes caused by low-level radiofrequency exposure, reported by Lin-Liu and Adey (1982), can most probably be explained by an increased diffusion. The increasing efficiency of various technological processes using heating by radiofrequency

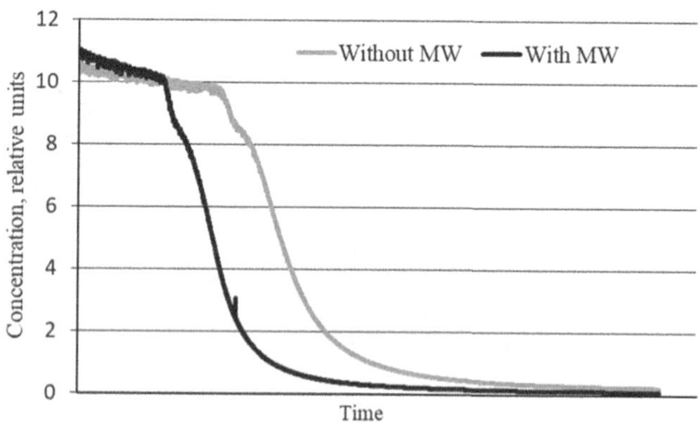

FIGURE 5.4 The measured salt concentrations with and without low-level microwave exposure at constant temperature.

radiation, reported by Liu et al. (2021), Wang et al. (2023), and Ranasinghe et al. (2025), has been explained by the authors as the effect of the radiofrequency radiation on diffusion.

5.7 EFFECT ON MEMBRANE POTENTIAL

Neurophysiologic processes are determined by the ions' movement due to electric forces and diffusion caused by a concentration gradient. The neurotransmitters' transit time in synapses and the balance of currents across the cell membrane depend on diffusion.

A balance between electric forces and diffusion occurs in a natural steady physiological condition. Alterations of the temperature of the medium affect diffusion, disturb the balance, and thus affect the majority of neurophysiologic processes. An external electric field causes electric polarization, in this way, affects electric forces in the medium, and disturbs the balance.

The Nernst–Planck equation is a continuity equation for ion concentration in a fluid medium. The ions' current density $\boldsymbol{J}$ is determined by diffusion due to an ionic concentration gradient ∇c and the electric field $\boldsymbol{E}$

$$\boldsymbol{J} = \boldsymbol{j}_D + \boldsymbol{j}_E = -D\nabla c + cq\boldsymbol{E} \tag{5.7}$$

where c is ion concentration, q is an ion charge, $\boldsymbol{j}_D$ is diffusion current density, and $\boldsymbol{j}_E = cq\boldsymbol{E}$ is electric current density. Nernst potential V across the cell membrane at equilibrium is determined by the balance between diffusion and electrical currents in the condition of zero current $\boldsymbol{J} = 0$, as follows

$$V = -\frac{D}{q}\ln\frac{c_i}{c_e} \tag{5.8}$$

where c_i is the intracellular concentration of the ions and c_e is the extracellular concentration of the ions.

The membrane potential increases linearly with diffusion.

Radiofrequency radiation enhances the diffusion rate and increases the diffusion coefficient D and the diffusion current. Increased diffusion current disturbs the balance between diffusion and electric field-produced currents in the Nernst–Planck equation (5.7) and finally alters the Nernst potential on the cell membrane (Eq. 5.8). This mechanism differs from the effect on membrane potential evaluated as unlikely by Adair (2003).

The alterations in membrane potential related to the radiofrequency radiation pulses can be considered similar to an event-related potential. The results reported by Bachmann et al. (2018) provide evidence of the detection of low-level microwave radiation-induced ERP in the recorded EEG signal.

The impact of microwave radiation on the concentration of various neurotransmitters in the brain has been reported in many independent studies (Hu et al., 2021). This knowledge and the impact of the radiation on membrane potential form the basis of microwave radiation effects in the brain's neurophysiology.

As demonstrated by Hinrikus et al. (2021), the positive results dominate the investigations of low-level radiofrequency radiation effects in human neurophysiology, indicated by EEG signals and behavior reported by many independent research groups.

5.8 IMPACT ON THE BRAIN'S NEURONAL SYSTEM

5.8.1 Effect on Brain Neuronal Oscillations

Neural activity in the brain can be described as a nonlinear oscillatory system (Section 4.3). Microwave radiation is a periodic oscillating process. The physical oscillations in one system can be coupled with the oscillations in another system. Physical oscillations in coupled systems are known to interact with each other and form a new picture of coupled oscillations. Hutcheon and Yarom concluded in 2000, based on the results reported by several researchers, that the brain's physiological oscillators act very similarly to electric oscillators.

Therefore, the neuronal oscillations in the brain are expected to be affected by external periodic electrical forces. The oscillating nature of the neuronal system is important for understanding the effect of radiofrequency radiation on brain functioning.

The discussions of the nature of brain neuronal oscillations by Nunez (1995), Steriade and Timofeev (2003), Buzsáki and Draguhn (2004), and Schnitzler and Gross (2005) contribute to understanding the behavior and functioning of the brain as a complex system of oscillations (Section 4.3). The coherent summation of oscillators in neuronal networks is an effective mechanism for the detection and amplification of weak signals. The neuronal oscillators behave as amplifiers active at the network level. The ability of neuronal assemblies to synchronize depends on the coupling strength and the distribution of natural frequencies.

A large number of nonlinear oscillators with linear all-to-all coupling and distributed natural frequencies form a complex nonlinear system (Matthews, 1991). The model for an individual component z_i of neuronal oscillations at frequency w_i is described by Matthews et al. (1991) as

$$\frac{dz_i}{dt} = \left(1 - \left|z_i\right|^2 + iw_i\right) z_i \qquad (5.9)$$

Equation (5.10) describes the system of N oscillators with coupling strengths K

$$\frac{dz_i}{dt} = \left(1 - \left|z_i\right|^2 + iw_i\right) z_i + \frac{K}{N} \sum_{i=j}^{N} \left(z_i - z_j\right) \qquad (5.10)$$

The numerical computer calculations of equation (5.10) were performed by varying the coupling strength and the width of the frequency distribution (Matthews et al., 1991). The modeling showed that the system exhibits extremely rich dynamics, including a variety of steady behaviors: frequency locking, amplitude death, incoherence, and partial locking. An unexpected result was that the system can exhibit unsteady behavior, in which the outcome distribution evolves periodically, quasi-periodically, or even chaotically.

The impact of microwave radiation can be considered an additional member in equation (5.10) at a frequency different from w_i. Such a modification of the governing equation unavoidably leads to different outcomes. A system of coupled neuronal oscillators can be affected by periodic external factors. The impact is expected to depend on the frequency of the affecting factor and its relation to neuronal oscillation frequencies. Since the microwave frequency is much higher than the frequencies of neuronal oscillations, a low-level modulation of the radiation is required to reveal the effect. The need for low-frequency modulation has been underlined first based on experimental results by Adey et al. (1982) and by many other researchers later. The dependence of the effect on modulation frequency is demonstrated by experiment (Hinrikus et al., 2008b). The effect decreases at modulation frequencies higher than 200 Hz and becomes insignificant at a modulation frequency of 1000 Hz.

The effect of low-level modulated microwave radiation on EEG rhythms has been reported by many independent research groups over decades (Hinrikus et al., 2021).

5.8.2 Impact on brain tissues

The structure of the brain contains intercellular and extracellular water. Neurons are embedded in an electrically conducting medium, the extracellular fluid, which, in principle, allows the extracellular activity of one cell to be perceived by its surrounding cells.

The effect of electric forces on brain fluid can be described by the general theory of electromagnetism. At the natural physiological state of the brain, the dipolar water molecules are located in the strong internal quasi-static electric field $\boldsymbol{E}_0$ created by bioelectric sources located outside and inside cells. The dipolar water molecules will tend to align themselves in the direction of the internal permanent or low-frequency bioelectric field. According to the formula (Eq. 1.11), the vector of the electric displacement field $\boldsymbol{D}_B$ in the brain medium

$$\boldsymbol{D}_B = \varepsilon_0 \left(1 + \chi_B\right) \boldsymbol{E}_0 \tag{5.11}$$

where $\boldsymbol{E}_0$ is the bioelectric field strength in a vacuum and $\chi_B \boldsymbol{E}_0$ presents polarization density, where ε_0 is the electric constant and χ_B is the electric susceptibility of the brain medium corresponding to the internal quasi-static bioelectric field.

The number of bioelectric sources inside the brain is constant. According to the Gauss law (Eq. 1.14), the total flow of the displacement field is also constant. From this condition, the electric field strength under the impact of internal biological fields $\boldsymbol{E}_B$ should be less compared to $\boldsymbol{E}_0$ and χ_B becomes negative

$$\boldsymbol{E}_B = \left(1 - \chi_B\right) \boldsymbol{E}_{B0} \tag{5.12}$$

The electric field of microwave radiation is weak compared to a strong bioelectric field inside the brain. In a linear medium, the impact of a weak microwave field is independent of the strong intercellular field (Section 1.11). The vector of the electric displacement field caused by the microwave radiation

$$\boldsymbol{D}_R = \varepsilon_0 \left(1 + \chi_R\right) \boldsymbol{E}_{R0} \tag{5.13}$$

where χ_R is the electric susceptibility of the medium corresponding to microwave radiation. The strength of the microwave electric field is much less compared to the internal quasi-static biological field. The electric permittivity decreases with frequency (Section 3.1). Therefore, the susceptibility at microwave frequencies is weaker compared to that at low frequency or static internal fields at biological frequencies, $\chi_R < \chi_B$. Therefore, the electric displacement field $\boldsymbol{D}_R \ll \boldsymbol{D}_B$. The impact of microwave radiation is much weaker compared to that of the internal biological field in a linear system.

The brain is a nonlinear system (Section 4.5). In a nonlinear medium, the mutual impact of two fields becomes evident as

$$\boldsymbol{D}_R = \varepsilon_0 \sqrt{\left(\boldsymbol{E}_R^2 + \boldsymbol{E}_B^2 + 2\boldsymbol{E}_R \boldsymbol{E}_B\right)} \tag{5.14}$$

In equation (5.14), the strong impact of the strong biological field dominates, as expected, compared to the microwave field. However, the third term indicates that the impact of the microwave field becomes remarkable. The impact of microwave radiation on the nonlinear brain is much stronger compared to that in a linear medium.

The radiofrequency radiation used in telecommunication systems is low-frequency modulated. Microwave radiation causes dielectric polarization of the tissues. Therefore, the dielectric polarization of brain fluid contains the components close to the frequencies of brain neuronal oscillations. The impact of the radiation not only on brain bioelectric oscillations but also on brain fluids can affect neurophysiologic processes and contribute to EEG signals.

The simulations performed by Anastassiou et al. (2010) and Popovych and Tass (2011) have demonstrated that local electric fields, generated by the cooperative action of brain cells, can influence the timing of neural activity and contribute to macroscopic field potential. If a neuronal population receives an external rhythmic input from an external electrical source, the population field potential may be entrained by the rhythmic input. The macroscopic field potential contributes to neuronal oscillations and EEG signals.

5.9 IMPORTANCE OF MODULATION

Low-frequency modulation of radiofrequency radiation is important in producing the effects of low-level radiation due to two main reasons. First, radiofrequencies are much higher physiological frequencies and cannot affect physiological processes directly. Second, human physiology can adapt to a permanent weak stressor, and the alteration in physiology can be repaired.

The information in radiofrequency systems is contained not in carrier radiofrequency but in low-frequency modulating signals. Modulated radiofrequency radiation is used in the majority of applications. The electric field of modulated radiofrequency radiation $\boldsymbol{E}_M$ can be presented as

$$\boldsymbol{E}_M = \boldsymbol{E}_0\left[1 - MF(t)\right]\cos(\omega t - kr) \tag{5.15}$$

where the modulation coefficient M describes the depth of modulation and function $F(t)$ is the time course of modulation that contains information.

Telecommunication systems use a pulse modulation where the parameters of the modulation are determined by the method of information coding, making the time course of the signals complicated. The frequency spectrum of modulation is determined by symbols' duration, timeslots, and frames corresponding to the utilized telecommunication standards (Table 5.1).

The numbers in Table 5.1 show that modulation frequencies in telecommunication systems are in the spectrum of physiological frequencies, lower than about 1000 Hz in all standards.

TABLE 5.1
Modulation frequency components in radiofrequency radiation related to the signal structure of telecommunication systems

Standard of telecommunication system	Symbol duration	Symbol rate	Time slot	Frame	Control, pilot, synchronization signals
2G – GSM	3.89 μs	270.833 ps	0.577 ms (1733 Hz)	5.616 ms (216.63 Hz) Multiframes 26 and 51 frames	216.68 Hz at the maximum power of the base station
3G – WCDMA	260 ns	3.84 Mps	0.666 ms (1501.5 Hz)	10 ms (100 Hz)	Continuous at 18% of base-station power
4G – LTE	66.7 μs	15,000 ps	1.0 ms (1000 Hz)	10 ms (100 Hz)	200 Hz 25 Hz
5G – NR FR1	66.7 μs	15,000 ps 30,000 ps	1.0 ms (1000 Hz) 0.5 ms (2000 Hz)	10 ms (100 Hz)	500 Hz 200 Hz 50 Hz
5G – NR FR2	66.7 μs	15,000 ps 30,000 ps 60,000 ps 120,000 ps	0.5 ms (2000 Hz) 0.25 ms (5000 Hz) 0.125 ms (8000 Hz) 0.0625 ms (16,000 Hz)	10 ms (100 Hz)	500 Hz 200 Hz 50 Hz
TETRA	3.92 μs	18,000 ps	15.167 ms (70.59 Hz)	56.67 ms (17.65 Hz)	17.63Hz At the maximum power of the base station
WiFi	3.2 μs	250,000 ps	4 ms (250 Hz)	40 ms (25 Hz)	10 Hz

The temporal character of modulation will be transferred to hydrogen bonds, diffusion, membrane potential, and other processes whose parameters are changed in time like *F(t)*. The fact that the level of the modulated electric field is less than thermal fluctuations is not critical. Even at a level much lower than thermal noise, the regular periodic electric field can excite harmonic oscillations while the stochastic thermal field cannot.

The functioning of human physiology is adapted to natural environmental conditions. A human organism can repair the alterations, restore normal functioning, and adapt to the changed conditions. Radiofrequency radiation alters the polarization of the brain tissue (Section 5.5) and therefore leads to the excitation of bioelectric oscillations in the brain (Section 5.6). The human brain can repair mild alterations and adapt to the low-level radiation.

The adaptation is an inertial, time-consuming process that can follow mild alterations caused by continuous-wave radiofrequency radiation. Contrariwise, the effect of periodic disturbance caused by the modulated radiofrequency radiation cannot be eliminated by adaptation or repair by the brain due to the inertia of the repairing process, which leads to periodic changes in the brain's bioelectric activity. The experiments by Hinrikus et al. (2008a) have demonstrated the ability of the brain to restore normal functioning, revealing the significant changes in the EEG signal only during the first 30 seconds of 60-second exposure half-cycle but not during the second half-cycle of 30–60 seconds.

The physiological response to microwave radiation becomes evident rather in the case of modulated radiation (Juutilainene et al., 2011). The importance of modulation in microwave effects in brain neurophysiology has been underlined by many authors, beginning with the first study. Bawin et al. reported that amplitude-modulated radiofrequency radiation-induced in vivo changes in calcium metabolism were not related to heating (Bawin et al. 1973). The experiments on calcium efflux by Lin-Liu and Adey (1982) and Adey et al. (1982) demonstrated that the effect is evident only with modulated radiation. The necessity of pulse modulation of microwave radiation to induce changes in waking and sleep EEG and cerebral blood flow is concluded by Huber et al. (2002, 2005). The effect of radiofrequency radiation on the brain would not be detectable without modulation of the radiation, and the modulation is crucial in revealing the effect.

The experimental data by Hinrikus et al. (2008a,b) support the dependence of the detected effect on modulation frequency. The changes detected in the EEG are larger at the modulation frequencies of 14 and 21 Hz, close to the frequencies of brain oscillations, and much smaller at the modulation frequencies of 70, 217, and 1000 Hz.

5.10 PARAMETRIC EXCITATION

The results of the experimental study reported by Hinrikus et al. (2008a,b) have led to an interesting finding that the effect of the modulated radiofrequency radiation on the human EEG signal varies depending on the modulation frequency. The alterations induced by the radiation are stronger in the EEG frequency bands, which are lower than the modulation frequency of the radiation. The effect of radiation has not been revealed in the EEG frequency bands higher than the modulation frequency.

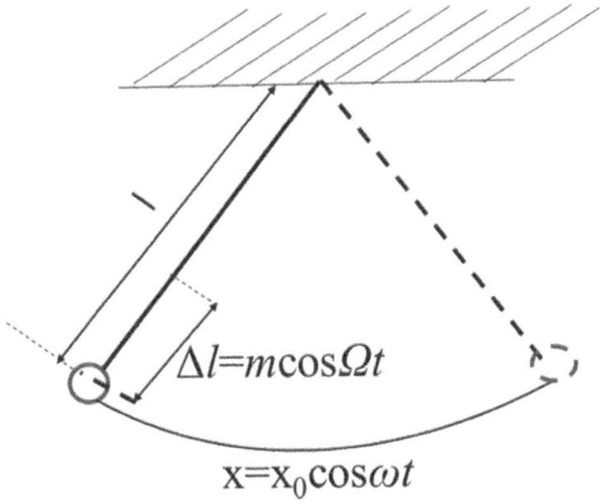

FIGURE 5.5 Parametric excitation of a pendulum, where l is the length of the pendulum, ω is the frequency of pendulum oscillations, and Δl indicates periodic excitation at the frequency of Ω.

The meander-shape pulse modulation used in the cited studies contains frequency components at the modulation frequency and higher harmonics but not at lower frequencies. The mystery of the detection at lower frequencies can be addressed only by parametric excitation of the oscillatory system. Parametric excitation is known to occur at a frequency lower than that of periodic excitation and can explain these experimental results.

A good example of parametric excitation is swinging. Periodic bending of the knees followed by a straightening out excites the swing oscillations. Such a movement of the knees is equivalent to periodic rising and lowering of the center of gravity of the body mass and so the length of the pendulum l (Figure 5.5). The swing motion starts from a small amplitude and increases during excitation due to energy from the source of periodic excitation. Parametric excitation occurs at an arbitrarily small initial disturbance, executing oscillations with progressively increasing amplitude.

The oscillatory nature of neuronal activity makes parametric excitation possible. The increase in EEG level at frequencies lower than the modulation frequency can most likely be explained by the parametric mechanism of excitation.

The theory of parametric excitation as "a theory of excitation of oscillations under the rather slow variations of a parameter" was proposed by Minorsky (1951). Based on the theory of parametric excitation, even a very small periodic force can result in the excitation of the oscillating system at fixed frequencies predetermined by the frequency of the periodic force. The parametric excitation of a linear system is described by the linear wave equation

$$\frac{d^2x}{dt^2} + 2\gamma\frac{dx}{dt} + \omega^2\left(1 - 2m\cos\Omega t\right) = 0 \quad (5.16)$$

where γ is the damping constant, ω is the frequency of the system, Ω is the frequency of parametric excitation, t is time, and m is the strength of excitation.

Tso and Caughey investigated in 1965 the theory of parametric excitation of a nonlinear system. The nonlinear wave equation describes a nonlinear system that is parametrically excited

$$\frac{d^2x}{dt^2} + 2\gamma\frac{dx}{dt} + x^{m1} + \beta x^{m2} + \omega^2(1 - 2m\cos\Omega t) = 0 \tag{5.17}$$

where γ is the damping constant, α and β are the nonlinearity constants at members $m1$ being odd and $m2$ being even integers, ω is the frequency of the system, Ω is the frequency of parametric excitation, t is time, and m is the strength of excitation. The damping factor γ is inversely proportional to the quality factor Q of the system. The threshold of parametric resonance occurs at

$$m = \frac{2\gamma}{\omega} = \frac{1}{Q} \tag{5.18}$$

Parametric resonance in a linear excitable system occurs under the condition

$$\frac{n\Omega}{2} = \omega \tag{5.19}$$

where n = 1, 2, 3, … is the zone number. In a linear excitable system, only one subharmonic of the excitation frequency $\Omega/2$ can exist.

Parametric excitation of a nonlinear excitable system has additional resonances and can occur under the condition

$$\frac{n\Omega}{4} = \omega \tag{5.20}$$

In a nonlinear excitable system, more than one subharmonic of the excitation frequency $\Omega/4$ can exist.

Parametric resonance is possible not only at the fixed resonance frequencies but also in the ranges of instability at frequencies lying on either side of their resonance values. As shown by Butikov (2004), the approximate expression for the boundaries $\omega_\pm$ of the zones of instability has been derived from the condition for the existence of a nontrivial solution of the linear equation wave equation (5.16)

$$\omega_\pm = 2\omega\left(1 \pm \frac{1}{2}\sqrt{m^2 - 1/Q^2} + \frac{11}{16}m^2\right) \tag{5.21}$$

The widths of the zones of instability ($\omega_+ - \omega_-$) and the strength of excitation m decrease with the zone number n. Therefore, the excitation is expected to be most effective inside the first two-three zones of instability.

Hinrikus et al. designed in 2011 a special experimental study to detect the possible parametric excitation of human EEG by modulated radiofrequency radiation using bands different from the traditional ones. The experiments were carried out using 450 MHz radiation at the field power density of 0.16 W/cm^2 modulated at 7, 14, 21, 40, and 70 Hz frequencies. Ten cycles, 1 minute of radiation on and

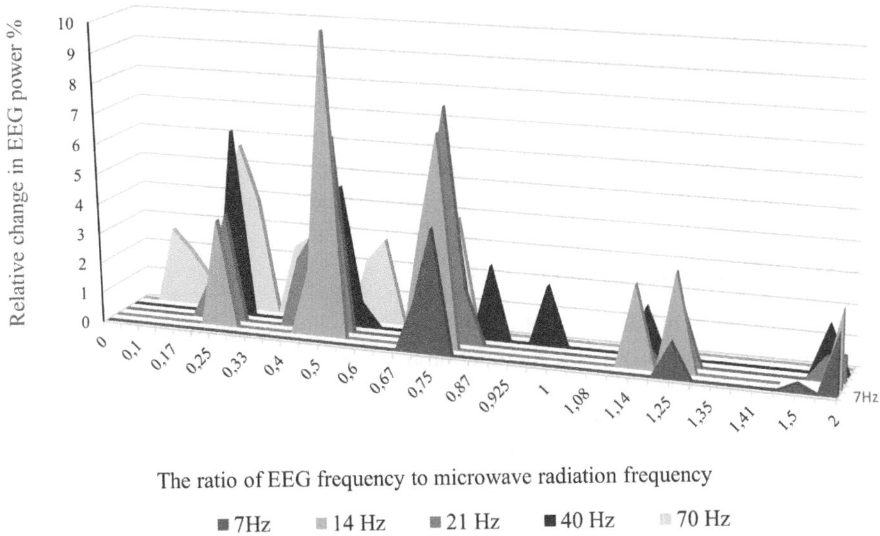

FIGURE 5.6 The relative changes in EEG power according to the ratio of EEG frequency to modulation frequency averaged over 14 subjects, 8 channels, and 10 cycles.

1 minute off, were performed at each modulation frequency. The powers of the resting EEG in 1 Hz bands at 15 fixed EEG frequencies, selected according to the modulation frequencies, were extracted from the EEG signal. The modulation frequencies $\omega = \Omega$ were filtered off. A relative change in the EEG power between sub-cycles with and without radiation was used as a quantitative measure to present the results (Figure 5.6).

The radiofrequency radiation increases the level of the EEG signal in the majority of selected analysis frequencies. However, the alterations in the EEG are statistically significant only at the ratio of EEG frequency to modulation $\omega/\Omega = 0.25$, $\omega/\Omega = 0.5$, and $\omega/\Omega = 0.75$. The maximal increase occurs at $\omega/\Omega = 0.5$, predicted by the linear theory of parametric excitation. The modulation frequency has been filtered off; therefore, the next expected great maximum at $\omega/\Omega = 1$ is missing. The excitation is evident not only at half but also at a quarter of the excitation frequency at $\omega/\Omega = 0.25$ and $\omega/\Omega = 0.75$, predicted by the nonlinear theory of parametric excitation. The parametric excitation at frequencies only within the first zones of instability $\omega/\Omega > 1$ can be explained by the theory of parametric excitation, predicting the decreasing efficiency of excitation with the increase of the zone number (Eq. 5.19). Additionally, fewer signals contribute to the higher zones of excitation because the frequencies 2 Ω are higher than the recorded EEG frequencies limited to 40 Hz. The results support the proposed mechanism of parametric excitation of neuronal activity in the brain by modulated radiofrequency radiation.

The parametric mechanism of excitation of the brain by modulated microwave radiation presumes maximal effect during the exposure to radiation and a decrease of the effect after the exposure caused by the dumping of the excited oscillations. The modulated microwave radiation has immediate and delayed but not long-lasting effects on the brain. However, living in an environment with modulated microwave radiation presents a permanent disturbance to the brain's bioelectric activity.

The experimental results show that not only subharmonics of order ½ predicted by the theory of parametric excitation in linear systems but also subharmonics of higher orders exist, predicted by the theory of parametric excitation of nonlinear systems. The results of the experiment demonstrate that the brain behaves as a nonlinear system regarding the response to exposure to radiofrequency radiation.

5.11 FREQUENCY LIMITS

The various mechanisms of radiofrequency radiation effects depend on the frequency of the radiation and also on the modulation frequency.

The limitations in radiofrequency can be caused by the inertia of particles that affect the electric permittivity and the impact on hydrogen bonds. The electric permittivity of tissues is known to decrease with frequency (Section 3.2). The frequency dependence of orientational polarization, the main component of electric permittivity at radiofrequencies, is limited to a frequency of about 300 GHz. The time constant of the dynamics of hydrogen bonds is about 1 ps (Section 5.3). The previous results confirm that there are no frequency limits on the radiation effects at radiofrequencies. However, the mechanism based on orientational polarization described in this book becomes small at frequencies higher than about 100 GHz. The mechanism of the effect at millimeter waves is still unknown.

The limitations of modulation frequency are determined by the physiological frequencies of living processes. The time constant of physiological processes is limited to about one millisecond. Therefore, the radiation modulated by a frequency higher than 1 kHz can be considered continuous. The experimental data reported by Hinrikus et al. (2008b) support the decrease of the effect on brain oscillations at the higher modulation frequencies.

To estimate the impact of frequency limitations in a living environment, it is important to know the modulation frequency components in radiofrequency radiation. The structure of the modulation signal is determined by information coding in the telecommunication system. The characteristic numbers for the most widely utilized telecommunication standards are presented in Table 5.1. The numbers in Table 5.1. show that the frequencies corresponding to the duration and rate of symbols are much higher than physiological frequencies. The frequencies corresponding to timeslot durations are in the range of physiological frequencies only for TETRA and WiFi. However, the frame durations, as well as control, pilot, and synchronization signals of all telecommunication systems, have components in the range of physiological signals. Therefore, the radiofrequency radiation utilized in telecommunication contains components that enable a possible impact on human physiological systems.

Parallel to telecommunication, broadcasting radio and TV also use modulated radiofrequency radiation. The carrier frequencies in broadcasting systems are lower than those used in mobile telecommunication. Radio and TV transfer audio signals that are in the range of physiological frequencies. Due to these reasons, the impact of broadcasting systems is expected to be even stronger compared to that of mobile telecommunication. The modulated radiofrequency radiation from broadcasting radio and TV cannot be neglected in discussing the possible impact of radiofrequency radiation on humans.

5.12 THRESHOLD OF NONTHERMAL EFFECT

Does the radiofrequency radiation have a threshold below which it does not have an effect? This is an important question and crucial to avoid possible consequences on health. The threshold of the radiofrequency radiation effects on living tissue can be determined by two factors: first, by the direct sensitivity of living cells to radiation, and, second, by the thresholds in physical mechanisms describing the effects

5.12.1 THE SENSITIVITY OF LIVING CELLS TO ELECTROMAGNETIC RADIATION

The sensitivity of a living cell as a receiver of radiofrequency radiation is limited by natural noise in the receiver. An approach similar to that in traditional radio engineering can be useful in estimating the sensitivity of living cells. The threshold of sensitivity for living tissue to radiation is determined by a fundamental physical noise in living tissues.

The physical noise in a living cell contains thermal noise, shot noise, and excess noise due to the opening and closing of ion channels. The noise related to random thermal fluctuations in electric potential has been predicted by Einstein, first experimentally demonstrated by Johnson, and interpreted by Nyquist. The level of thermal Johnson noise at the resistance R is determined by the Nyquist formula

$$(\delta U)_T^2 = 4RkT\Delta f \tag{5.22}$$

where $k = 1.38 \times 10^{-23}$ [J/K] is the Boltzmann constant, T [K] is the absolute temperature, and Δf is the frequency band.

Shot noise is caused by the discrete nature of current carriers, ions, or electrons. The random numbers of charged particles per unit time cause current fluctuations. The random fluctuations in current were first studied by Schottky. The Schottky formula determines the random fluctuation of the current I

$$(\delta I)_I^2 = 2eI\Delta f \tag{5.23}$$

where $e = 1.6 \times 10^{-19}$ C is the electron charge.

Excess noise is related to the opening and closing of membrane channels. The random nature of the opening and closing of channels results in additional fluctuations of currents through the membrane. The total current through a cell membrane can be considered the average of single-channel currents over all channels. Due to the random nature of opening–closing independent channels, the fluctuations in the total current through the cell membrane, in the case of large numbers of channels, are much less than the excess noise in a single channel.

A physical single-cell model can be considered as a possible receiver of the electric field. The thermal noise limit was considered by Weaver and Astumian (1990).

The level of thermal noise was estimated based on the model of a spherical cell of 10 μm radius at $T = 310$ K. However, the value of Δf was unpredictable due to unknown membrane properties. The model presumes the theoretical value of 10^{-5}–10^{-4} V/m for thermal noise on a cellular membrane at $\Delta f = 10$ Hz. This limit corresponds to the noise power spectral density of about 10^{-8} W/m^2.

The level of total noise, including all three components, thermal, current, and excess noises in living cells, was estimated by Hinrikus et al. (1998). The spectral density of noise was calculated on a cell model with typical cell parameters: membrane resistance $R = 1$ M, single-channel current $I = 1$ pA, and number of channels 10^5. The performed calculations result in a value of total noise spectral density on a cell membrane of 10^{-7}–10^{-8} W/m^2. This result is in good agreement with the numbers reported by Weaver and Astumian for thermal noise. However, the cumulative effect of coherent radiofrequency radiation can change the plausible threshold (Section 5.5) much lower than that determined by thermal noise. The impact of integration during exposure time was discussed by Weaver and Astumian (1990). They demonstrated the decrease of the minimum detectable field with the duration of exposure up to several orders of magnitude. Hinrikus et al. discussed in 2007 the integration of small differences as a possibility to detect signals of much lower power than the level of random noise. Coherent radiofrequency radiation is expected to be sensed by the cell even if its power is much lower than the level of random thermal noise.

5.12.2 The Threshold of Physiological Effects

The thermal mechanism of the effects of radiofrequency radiation is described by the Arrhenius model (Section 5.1). The Arrhenius model has no principal threshold. Even a very small increase in temperature can affect human physiology.

The plausibility of non-thermal effects was evaluated by Adair (2003) using the thermodynamic criterion that is based on the comparison with the level of thermal radiation.

An electric polarization is the origin of the nonthermal mechanisms of the effects of radiofrequency radiation. The electric polarization is linearly related to the electric field in the electric susceptibility (Eq. 1.10). There are no theoretical or experimental data about the threshold of electric polarization. The linear relationship between electric field and polarization presumes a constant value of electric susceptibility. There is a possibility that the susceptibility to the electric field can be disturbed by the thermal random fluctuations of energy kT, much higher than the energy of radiofrequency radiation. The cumulative effect of coherent radiofrequency radiation discussed above decreases the threshold significantly lower than the thermal limit. This conclusion is supported by experimental data of measurements of electric permittivity produced by radiofrequency radiation despite the $hf < kT$ (Section 3.1).

The restructuring and weakening of hydrogen bonds between dipolar molecules by radiofrequency radiation is related to the synchronous rotation of molecules. As in the case of electric polarization, the kT limit does not determine the threshold of the alterations in hydrogen bonds. Experimental data reported by Fesenko and Gluvstein (1995) have demonstrated that low-level microwave radiation alters the properties of water. This result supports the restructuring of hydrogen bonds by radiation. There is no data about the threshold of restructuring the intramolecular bonds.

The theory of parametric excitation predicts a threshold determined by the damping factor and losses in the system. In the case of neuronal oscillations in the brain, biochemical energy can compensate for the losses. Therefore, the threshold of parametric excitation is not defined.

Hinrikus et al. (2021) have formulated the conclusion that there are no data about the factors determining the threshold of the low-level radiofrequency effect. The "linear, no-threshold" model is also used in low-level ionizing radiation, which assumes that there is no dose so small that it has no effect.

5.13 MODEL OF PHYSICAL MECHANISM

In previous sections of this chapter, all physical phenomena describing the effect of radiofrequency radiation on humans have been discussed in detail. The preceding discussion has shown that the factors contributing to the non-thermal effect mechanism have no limitations on the frequency or level of radiation in the microwave band. These conclusions lead to a physical model of the radiofrequency radiation effect first described by Hinrikus et al. (2018).

Figure 5.7 presents the upgraded structure of the model. The origin of the effects, dielectric polarization, is common in neurophysiological and biological tracks. Dielectric polarization is a linear process regarding electric field strength (Eq. 1.10). Dielectric polarization caused by the modulated radiofrequency radiation can excite oscillations in brain tissues. The restructuring of hydrogen bonds caused by dielectric polarization can lead either to increased diffusion and neurophysiological effects or to the instability of molecules, oxidative stress, and biological effects. The electric forces affecting dipolar molecules are linear (Eq. 1.3).

The first track presents a sequence of processes beginning from the excitation of brain tissues (Section 5.7) and increased diffusion (Section 5.5), followed by alterations in membrane potential (Section 5.6), and leads to excited neuronal oscillations and disturbance of neuronal oscillations in the brain (Section 5.8). Finally, an impact on neurophysiological or mental disorders is possible if the recovery mechanisms of the living organism are insufficient to repair the changes made by radiation. No alterations in cellular structure are evident. All steps of the first track in the model can be explained based on the physical laws and described by relevant formulas.

Living tissues can be considered as a linear medium regarding the processes of radiofrequency wave propagation and linear effects like dielectric polarization. The linear theory of electromagnetism is applicable, considering tissues as a linear medium. However, as demonstrated by parametric excitation, the brain behaves as a nonlinear system regarding the physiological response of living organisms to radiation. Discussing the impact of radiation on physiological and biological processes like nerve pulse propagation or membrane potential, a nonlinear model should be used. In the case of neuronal systems, the nonlinear Hodgkin–Huxley equations describe the excitation of neurons. The similarity between neuronal oscillations in the brain and physical oscillations allows us to use the model knowledge from the oscillations theory for describing neuronal oscillations (Balzano and Sheppard, 2003).

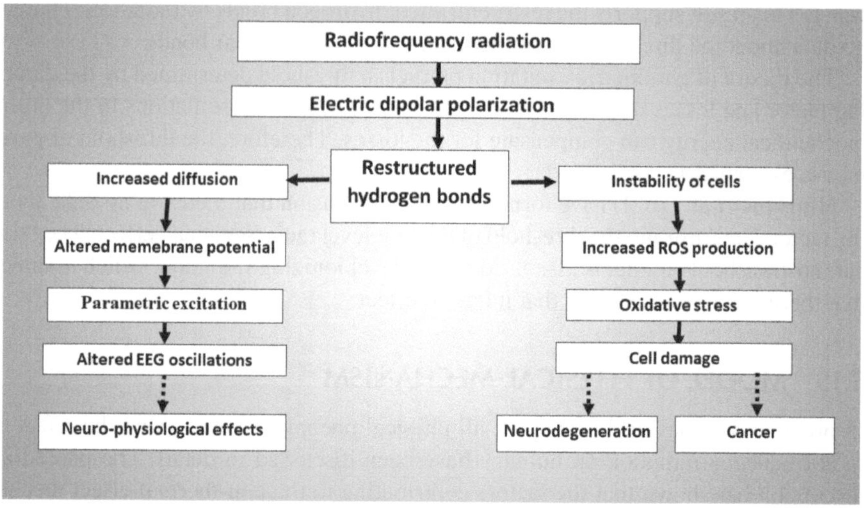

FIGURE 5.7 The model of a mechanism of microwave frequency radiation effects on the brain.

The steps in the model, beginning from the restructuring of hydrogen bonds, can be considered nonlinear. A very weak initial alteration on the neuronal or molecular level can lead to unpredictable consequences on health – or not cause any remarkable health effect. Reversible health effects are caused by alterations in physiological processes (first track in Figure 4.7). Irreversible health damage like cancer is caused by changes in DNA and tissues (second track in Figure 4.7).

The second track starts from the instability of cells caused by the weakening intermolecular hydrogen bonding. Weakening intermolecular bonds can increase the level of reactive oxygen species (ROS), followed by increased oxidative stress (Section 5.4).

The second track can lead to alterations of cellular structure and cell damage. Biological systems are complex and nonlinear, and the state of a system depends on a very large number of parameters. There are still no mathematical equations describing the impact of external factors on a biological system. The effects caused by radiation can only be supported by experimental data. Further consequences in health, including genome instability and neurodegeneration, are possible.

Finally, both tracks, neurophysiological and biological, can lead to health effects. Whereas compensation mechanisms by living organisms can repair the temporal changes in brain neurophysiology and the ROS formation process, the health effects do not necessarily become evident. The outcome of the transformation of biological effects to impact health in nonlinear biological systems is unpredictable. In a nonlinear system, even very small alterations in initial or boundary conditions can cause a big change in the result. Due to the complex character of transferring the changes to a health disorder, the outcome is unpredictable (Chapter 4).

5.14 CONCLUSIONS

This chapter presents the physical mechanism of coherent microwave radiation effects. There is no need to use the concept of a new special mechanism to understand the impact of microwave radiation. Dielectric polarization, a well-known and commonly accepted phenomenon, can trigger small initial alterations sufficient to result in significant changes in nonlinear living systems.

- Dielectric polarization is the only non-thermal physical mechanism of electromagnetic radiation that is commonly accepted and has no frequency restrictions at microwaves.
- Due to the impact of coherent cumulative summarizing, low-level coherent microwave radiation can disturb the mutual structure of dipolar water molecules.
- Living systems are nonlinear. Even a small alteration in the initial condition can result in a significant change in the outcome.
- The energy producing the microwave radiation effect is coming not from the radiation but from the metabolism supporting physiological processes affected due to small alterations in initial conditions. The nonthermal effect has no principal physical threshold.
- The nonthermal effect is produced by the electric field strength and increases linearly with the electric field, not the radiation power density and SAR.
- Low-level microwave radiation is a weak physical stressor. Due to the complexity and nonlinearity of living systems, the biological responses to the radiation are diverse and unpredictable.

REFERENCES

Adair RK, (1999). Effects of very weak magnetic fields on radical pair reformation. *Bioelectromagnetics*. 20(4):255–263.

Adair RK, (2002) Vibrational resonances in biological systems at microwave frequencies. *Biophys J*. 82: 1147–1152.

Adair RK, (2003) Biophysical limits on a thermal effects of RF and microwave radiation. *Bioelectromagnetics*. 24(1):39–48.

Adey WR, Bawin SM, Lawrence AF, (1982). Effects of weak amplitude-modulated microwave fields on calcium efflux from awake cat cerebral cortex. *Bioelectromagnetics*. 3(3):295–307.

Aly AA, Cheema MI, Tambawala M, Laterza R, Zhou E, Rathnabharathi K, Barnes FS, (2008). Effects of 900-MHz radio frequencies on the chemotaxis of human neutrophils in vitro. *IEEE Trans Biomed Eng*. 55(2 Pt 1):795–797. doi: 10.1109/TBME.2007.912636.

Anastassiou CA, Montgomery SM, Barahona M, Buzsáki G, Koch C, (2010). The effect of spatially inhomogeneous extracellular electric fields on neurons. *J Neurosci*. 30:1925–1936.

Antonio C, Deam RT, (2007). Can "microwave effects" be explained by enhanced diffusion? *Phys Chem Chem Phys*. 9: 2976–2982.

Apollonio F, Liberti M, Paffi A, Merla C, Marracino P, Denzi A, (2013). Feasibility for microwaves energy to affect biological systems via nonthermal mechanisms: a systematic approach. *IEEE Trans Microw Theory Tech*. 61(5):2031–2045.

Aronica R, Enrico P, Squarcina L, Brambilla P, Delvecchio G, (2022). Association between diffusion tensor iImaging, inflammation and immunological alterations in unipolar and bipolar depression: A review. *Neurosci Biobehav Rev.* 143:104922. doi: 10.1016/j.neubiorev.2022.104922.

Astumian RD, (2003). Adiabatic pumping mechanism for ion motive ATPases. *Phys Rev Letts.* 91:118102 1–4.

Bachmann M, Päeske L, Ioannides AA, Lass J, Hinrikus H, (2018). After-effect induced by microwave radiation in human electroencephalographic signal: a feasibility study. *Int J Radiat Biol.* 94(10):896–901.

Bakker J, Woutersen S, Nienhuys H-K, (2000). Reorientational motion and hydrogen-bond stretching dynamics in liquid water. *Chem Phys.* 258:233–245.

Balzano Q, Sheppard A (2003). RF nonlinear interactions in living cells-I: nonequilibrium thermodynamic theory. *Bioelectromagnetics.* 24(7):473–482. doi: 10.1002/bem.10126. Erratum in: *Bioelectromagnetics.* 2007 Jan;28(1):47.

Balzano Q, Hodzic V, Gammon RW, Davis CC, (2008). A doubly resonant cavity for detection of RF demodulation by living cells. *Bioelectromagnetics.* 29:81–91.

Barnes FS, (1992). Some engineering models for interactions of electric and magnetic fields with biological systems. *Bioelectromagnetics.* Suppl 1:67–85. doi: 10.1002/bem.2250130708.

Barnes F, Greenebaum B, (2015). The effects of weak magnetic fields on radical pairs. *Bioelectromagnetics.* 36(1):45–55.

Barnes F, Greenebaum B, editors. (2018). *Biological and medical aspects of electromagnetic fields.* CRC Press, Taylor & Frances, Boca Raton, FL.

Barnes F, Greenebaum B, (2020). Possible mechanism for synchronized detection of weak magnetic fields by nerve cells. *Bioelectromagnetics.* 41(3):213–218.

Bawin SM, Gavalas-Medici RJ, Adey WR, (1973). Effects of modulated very high frequency fields on specific brain rhythms in cats. *Brain Res.* 30;58(2):365–384.

Bawin SM, Kaczmarek LK, Adey WR, (1975). Effects of modulated VHF fields on the central nervous system. *Ann N Y Acad Sci.* Feb 28;247:74–81.

Bohr H, Bohr J, (2000). Microwave-enhanced folding and denaturation of globular proteins. *Phys Rev.* E61:4310–4314.

Buzsáki G, Draguhn A, (2004). Neuronal oscillations in cortical networks. *Science.* 304(5679):1926–1929. doi: 10.1126/science.1099745.

Butikov EI, (2005). Parametric excitation of a linear oscillator. *Eur J Phys.* 25:535–555.

Buzsáki G, Anastassiou CA, Koch C, (2012). The origin of extracellular fields and currents--EEG, ECoG, LFP and spikes. *Nat Rev Neurosci.* May 18;13(6):407–420.

Challis LJ, (2005). Mechanisms for interaction between RF fields and biological tissue. *Bioelectromagnetics.* Suppl 7:S98–S106. doi: 10.1002/bem.20119.

Chiabrera A, Bianco B, Moggia E, Kaufman JJ, (2000). Zeeman Stark modelling of the RF EMF interaction with ligand binding. *Bioelectromagnetics.* 21:312–324.

Dill KA, Bromberg A, (2002). *Molecular driving forces: statistical thermodynamics in chemistry and biology.* Carland Science, New York, London.

El Hajj W, El Khatib N, Volpert V (2023). Inflammation propagation modeled as a reaction-diffusion wave. *Math Biosci.* 365:109075. doi: 10.1016/j.mbs.2023.109075.

Fesenko EE, Gluvstein AY, (1995). Changes in the state of water, induced by radiofrequency electromagnetic fields. *FEBS Letters.* 367:53–55.

Foster KR, (2000). Thermal and nonthermal mechanisms of interaction of radio-frequency energy with biological systems. *IEEE Trans Plasma Sci.* 28(1):15–23.

Frölich H, (1968). Long range coherence and energy storage in biological systems. *Int J Quant Chem.* 2:641–649.

Galema SA, (1997). Microwave chemistry. *Chem Soc Rev.* 26:233–238.

Gilli P, Gilli G, (2009). *The nature of the hydrogen bond.* Oxford University Press, Oxford.

Henshaw DL, Philips A (2025). A mechanistic understanding of human magnetoreception validates the phenomenon of electromagnetic hypersensitivity (EHS). *Int J Radiat Biol.* 101(2):186–204.

Hinrikus H, Lass J, Riipulk J, (1998).The sensitivity of living tissue to microwave field. *Proceedings of the 20th Annual International Conference of the IEEE Engineering in Medicine and Biology Society* 20:3249–3252, 29 Oct.-1 Nov 98, Hong Kong.

Hinrikus H, Bachmann M, Tomson R, Lass J, (2005). Non-thermal effect of microwave radiation on human brain. *Environmentalist.* 25:187–194.

Hinrikus H, Bachmann M, Kalda J, Sakki M, Lass J, Tomson R, (2007). Methods of electroencephalographic signal analysis for detection of small hidden changes. *Nonlinear Biomed Phys.* 1(1):9. doi: 10.1186/1753–4631-1–9.

Hinrikus H, Bachmann M, Lass J, Tomson R, Tuulik V, (2008a). Effect of 7, 14 and 21 Hz modulated 450 MHz microwave radiation on human electroencephalographic rhythms. *Int J Rad Biol.* 84:69–79.

Hinrikus H, Bachmann M, Lass J, Karai D, Tuulik V, (2008b). Effect of low frequency modulated microwave exposure on human EEG: individual sensitivity. *Bioelectromagnetics.* 29(7):527–538.

Hinrikus H, Bachmann M, Lass J, (2011). Parametric mechanism of excitation of the electroencephalographic rhythms by modulated microwave radiation. *Int J Radiat Biol.* 87(11):1077–1085.

Hinrikus H, Lass J, Karai D, Pilt K, Bachmann M, (2015). Microwave effect on diffusion: a possible mechanism for non-thermal effect. *Electromagn Biol Med.* 34(5):327–333.

Hinrikus H, Bachmann M, Karai D, Lass J, (2017). Mechanism of low-level microwave radiation effect on nervous system. *Electromagn Biol Med.* 36(2):202–212.

Hinrikus H, Bachmann M, Lass J, (2018). Understanding physical mechanism of low-level microwave radiation effect. *Int J Radiat Biol.* 94(10):877–882.

Hinrikus H, Lass J, Bachmann M, (2021). Threshold of radiofrequency electromagnetic field effect on human brain. *Int J Radiat Biol.* 97(11):1505–1515.

Hoz A, Díaz-Ortiz A, Moreno A, (2005). Microwaves in organic synthesis. Thermal and non-thermal microwave effects. *Chem Soc Rev.* 34:164–178.

Hu C, Zuo H, Li Y, (2021). Effects of radiofrequency electromagnetic radiation on neurotransmitters in the brain. *Front Public Health.* 9:691880. doi: 10.3389/fpubh.2021.691880.

Huang JY, Liu T, Zhang YM, Hu PC, (2025). Comparative study of microwave-assisted and hydrothermal hydroxyapatite synthesis from neutralization slag: optimization, energy, and adsorption. *Ceram Int.* 50I(21):42157–42168 PartB.

Huber R, Treyer V, Borbély AA, Schuderer J, Gottselig JM, Landolt HP, Werth E, Berthold T, Kuster N, Buck A, Achermann P, (2002). Electromagnetic fields, such as those from mobile phones, alter regional cerebral blood flow and sleep and waking EEG. *J Sleep Res.* 11(5):289–295.

Huber R, Treyer V, Schuderer J, Berthold T, Buck A, Kuster N, Landolt HP, Achermann P, (2005). Exposure to pulse-modulated radio frequency electromagnetic fields affects regional cerebral blood flow. *Eur J Neurosci.* 21(5):1000–1006.

Hutcheon B, Yarom Y, (2000). Resonance, oscillation and the intrinsic frequency preferences of neurons. *Trends Neurosci.* 23:216–222.

Juutilainen J, Höytö A, Kumlin T, Naarala J, (2011). Review of possible modulation-dependent biological effects of radiofrequency fields. *Bioelectromagnetics.* Oct;32(7):511–534.

Kowalczuk C, Yarwood G, Blackwell R, Priestner M, Sienkiewicz Z, Bouffler S, Ahmed I, Abd-Alhameed R, Excell P, Hodzic V, Davis C, Gammon R, Balzano Q, (2010). Absence of nonlinear responses in cells and tissues exposed to RF energy at mobile phone frequencies using a doubly resonant cavity. *Bioelectromagnetics.* 31:556–565.

Laurence JA, French PW, Lindner RA, McKenzie DR, (2000). Biological effects of electromagnetic fields-mechanisms for the effects of pulsed microwave radiation on protein conformation. *J Theor Bio*. 206:291–298.

Laurence JA, McKenzie DR, Foster KR (2003). Application of the heat equation to the calculation of temperature rises from pulsed microwave exposure. *J Theor Biol*. 222:403–405.

Lin-Liu S, Adey WR, (1982) Low-frequency amplitude modulated microwave fields change calcium efflux rates from synaptosomes. *Bioelectromagnetics*. 3:309–322.

Liu Q, He, WQ, Aguedo, M, Xia X, Bai WB, Dong YY, Song JQ Richel A, Goffin D, (2021). Microwave-assisted alkali hydrolysis for cellulose isolation from wheat straw: influence of reaction conditions and non-thermal effects of microwave. *Carbohydr Polym*. 253: 117170. doi: 10.1016/j.carbpol.2020.117170

Luzar A, Chandler D, (1996). Hydrogen-bond kinetics in liquid water. *Nature*. 379:55–57.

Matthews PC, Mirollo RE, Strogatz SH, (1991). Dynamics of a large system of coupled nonlinear oscillators. *Phys. D: Nonlinear Phenom*. 52(2–3):293–331.

Meyer F, Bitsch A, Forman HJ, Fragoulis A, Ghezzi P, Henschenmacher B, Kellner R, Kuhne J, Ludwig T, Sachno D, Schmid G, Tsaioun K, Verbeek J, Wright R, (2024) The effects of radiofrequency electromagnetic field exposure on biomarkers of oxidative stress in vivo and in vitro: a systematic review of experimental studies. *Environ Int*. Aug 14;194:108940. doi: 10.1016/j.envint.2024.108940.

Minorsky N, (1951). Parametric excitation. *J. Appl. Phys*. 11(1):49–54.

Nunez PL, (1995). *Neocortical dynamics and human EEG rhythms*. Oxford University Press, New York.

Pearce J, (2018). Irreversible tissue thermal alterations: skin burns, thermal damage and cell death. In: D. Shrivastava, editor. *Theory applications of heat transfer in humans*, vol. 2. Wiley, Hoboken, NJ, pp. 553–590.

Polk C, Postow, E, editors, (1986). *Handbook of biological effects of electromagnetic field*. CRC Press, Boca Raton, FL.

Popovych OV, Tass PA, (2011). Macroscopic entrainment of periodically forced oscillatory ensembles. *Prog Biophys Mol Biol*. 105:98–108.

Ranasinghe M, Sivapragasam N, Mostafa H, Airouyuwa JO, Manikas, Sundarakani B, Maqsood S, Stathopoulos C, (2025). Valorizing date seeds in biscuits: a novel approach to incorporate bioactive components extracted from date seeds using microwave-assisted extraction. *Resour Environ Sustain*. 15:100147. doi: 10.1016/j.resenv.2023.100147

Schmidt R, Thews G, (1983). *Human physiology*. Springer, Berlin.

Schnitzler A, Gross J, (2005). Normal and pathological oscillatory communication in the brain. *Nat Rev Neurosci*. 6: 285–296.

Schuermann D, Mevissen M, (2021). Manmade electromagnetic fields and oxidative stress-biological effects and consequences for health. *Int J Mol Sci*. 22(7):3772. doi: 10.3390/ijms22073772.

Schwan HP, (1985). EM-field induced force effects In: A Chiabrera, C Nicolini, HP Schwan, editors. *Interactions between electromagnetic fields and cells*. Plenum Publishing Corp, New York.

Sheppard AR, Swicord ML, Balzano Q, (2008). Quantitative evaluations of mechanisms of radiofrequency interactions with biological molecules and processes. *Health Phys*. 95(4):365–396.

Shiraga K, Adachi A, Nakamura M, Tajima T, Ajito K, Ogawa Y, (2017). Characterization of the hydrogen-bond network of water around sucrose and trehalose: microwave and terahertz spectroscopic study. *J Chem Phys*. 146:105102.

Silberberg MS, (2012). *Chemistry: the molecular nature of matter and change*. McGraw-Hill, New York.

Steriade M, Timofeev I, (2003). Neuronal plasticity in thalamocortical networks during sleep and waking oscillations. *Neuron*. 37:563–576.

Tso WK, Caughey TK, (1965). Parametric excitation of a nonlinear system. *J Appl Mech.* 32: 899–902.

Wang NN, Zou WH, Li XY, Liang YQ, Wang, P, (2023). Study and application status of the nonthermal effects of microwaves in chemistry and materials science - a brief review. *RSC Advances.* 12: 17158–17181.

Weaver JC, Astumian RD, (1990). The Response of living cells to very weak electric fields: the thermal noise limit. *Science.* 247:459–462.

Yakymenko I, Tsybulin O, Sidorik E, Henshel D, Kyrylenko O, Kyrylenko S, (2016). Oxidative mechanisms of biological activity of low-intensity radiofrequency radiation. *Electromagn Biol Med.* 35(2):186–202.

Zahn M, (2003). *Electromagnetic field theory: a problem solving approach.* John Wiley & Sons, New York.

Zong L, Zhou S, Sgriccia N, Hawley MC, Kempel LC, (2003). A review of microwave-assisted polymer chemistry (MAPC). *J Microw Power Electromagn Energy.* 38(1):49–74.

6 Effect of microwave radiation on the human brain

HIGHLIGHTS

- The majority of experimental studies on microwave effects on the human brain have reported positive results.
- The diversity between methods and parameters in the performed studies does not allow us to evaluate the characteristics of the radiation effects.
- The developed advanced method of integration of differences enables the suppression of the EEG signal's natural fluctuations and can detect information about the features of the microwave effect on the human brain which is hidden in traditional methods.
- The low-level microwave effect on the brain depends on the modulation frequency and radiation level but not significantly on the carrier frequency.
- The effect is proportional to the electric field strength rather than the specific absorption rate.
- The effect differs between individuals. Only a minority, around 30% of exposed participants, were significantly affected.
- The difference appears in the EEG signal, which directly reflects brain activity, rather than in behavior and sleep quality.
- The response to modulated microwave radiation pulses in the EEG signal is similar to the visual evoked potential.

6.1 INTRODUCTION

The laws of electromagnetism expressed in mathematical form have been the basis of the mechanisms of the radiofrequency radiation effects, and the performed experiments only supported the validity of the conclusions delivered in the previous chapters. The propagation of radiofrequency radiation in living tissues can be discussed using a linear approach, assuming the tissues behave as a linear medium. The transition from a physical to a biological–physiological approach requires replacing a linear model with a nonlinear chaotic model. Due to the huge number of variables and the chaotic nature of the processes in living systems, mathematical modeling is hardly possible. Therefore, experimental studies have become the basis for discussing the effects of radiofrequency radiation on the human brain.

DOI: 10.1201/9781003729976-8

The effect of microwave radiation on has been investigated by many research groups for decades (Valentini et al., 2007; Kwon and Hämalainen, 2011, Hinrikus et al., 2022, Lai and Levitt, 2022). The main attention has been on low-level effects possible in everyday living conditions. Despite the long-term, wide-scale investigations, there is very limited knowledge about the features of the effects. The main reason is that the effects are weak and veiled by natural changes in our physiology related to various possible factors. The long-term alterations are caused by altered environmental conditions, nutrition, noise levels, etc. The short-term alterations in the EEG signal are hidden by the huge number of signals related to brain functioning. The alterations in the brain caused by low-level radiofrequency radiation are mild and difficult to detect.

Therefore, sensitive methods are required to reveal the radiofrequency radiation-related changes in the brain. On the contrary, the method should be passive, not disturbing the brain functioning. Most brain medical imaging methods utilizing radiation and chemicals appear to be unsuitable in the current case. An electroencephalographic (EEG) method, being passive, sensitive, and providing a good temporal resolution, is widely used in studies on radiofrequency radiation effects on the human brain. The author's research group has developed a method for reducing the impact of natural fluctuations in the EEG signal that provides higher sensitivity and enables more detailed information about the features of the microwave effect on the brain.

This chapter concentrates on the analysis of experimental results on low-level microwave effects on the brain. The chapter includes an analytical review of published experimental results and determines the nature of the effects based on the results of performed investigations achieved by applying the developed novel method and the knowledge presented in previous chapters.

6.2 ANALYSIS OF EXPERIMENTAL RESULTS ON NEUROPHYSIOLOGICAL EFFECTS IN THE HUMAN BRAIN

Over decades, hundreds of EEG studies on low-level radiofrequency effects on the human brain have been performed (Hinrikus et al., 2022; Lai and Levitt, 2022). Despite that, doubts about the non-thermal effects are still expressed, and many questions have not yet been exhaustively answered.

The author has collected the database of the human studies on the effects of resting-state EEG, sleep EEG, event-related potential (ERP), cognition, and behavior published in peer-reviewed journals over about 20 years. The analysis of my database 2007–2025 demonstrates some trends in the investigation of effects. The studies on EEG have been limited mainly to 2.4 GHz frequencies; only one human study on 3.6 GHz modulated by 5G signals by Sousouri et al. (2025) has been recently published. Of the 77 selected for the EEG studies, 61% (47 studies) have reported alteration in the investigated parameters with radiofrequency radiation. Figure 6.1 presents the results of the distribution of the studies according to the method used and positive–negative results in detecting the effect of radiofrequency radiation. Two studies

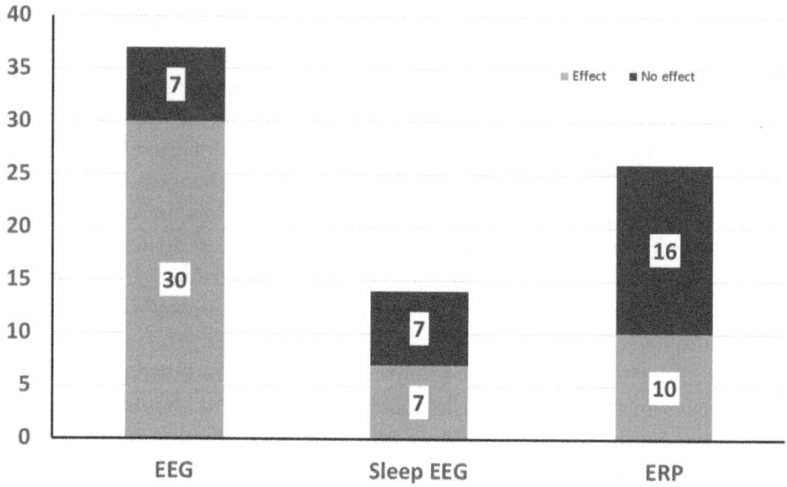

FIGURE 6.1 The number of positive and negative results reported in EEG; sleep EEG; and ERP, cognition, and behavior radiofrequency radiation studies, 2007–2024.

by Curcio et al. (2009) and Volkow et al. (2011) on brain hemodynamics and metabolism have indicated positive results.

The largest number of studies, 37, has been performed using resting-state EEG; 26 using ERPs, cognition, and behavior; and 14 using sleep EEG. The relative part of positive findings indicating the effect of radiation is highest, 81% for resting EEG, followed by 50% for sleep EEG, and 38.5% for ERP, cognition, and behavior. These numbers are consistent with the results of previous analysis performed on a somewhat more limited database by Hinrikus et al. (2021) over the years from 2007 to 2020.

The distribution of sensitivity with different methods can be explained by two possible reasons. First, the character of signals recorded in the studies differs when utilizing different methods. The brain receives only a little information in the resting-state, and the eyes closed EEG signal contains a minimum level of confounding factors. The minimal level of additional alterations increases the possibility of detecting the impact of the applied radiofrequency radiation. The processes of cognition are more complicated, containing the additional steps of reception and processing of external information. The sleep EEG reflects, in addition to the radiation effects, the process of generation of natural spindles in the sleep structure that decreases the possibility of detection of the direct effect of radiation. Second, the nature of the effects is different. The impact of the brain neuronal oscillation affected by the radiation can be repaired at the cognitive and behavioral levels.

Tables 6.1–6.3 present the distribution of studies according to detected effects separately in three categories of methods: resting EEG; sleep EEG; and ERP, cognition, and behavior. The rationale for a separate discussion of the results detected by different methods is the different physiological nature of the effects.

TABLE 6.1
Distribution of resting-state EEG studies according to the reported radiofrequency radiation effects

Effect	Frequency GHz	Modulation	SAR (W/kg)	References
Increased theta	0.88–0.94	PM		Bardasano et al. (2007)
Increased alpha	1.9; 0.94; 0.92; 0.902; 0.9; 0.89; 0.88; 0.45	GSM WCDMA; LTE PM: 7; 14; 21; 40; 70; 217 Hz	2; 1.18; 1; 0.93; 0.955; 0.69; 0.674; 0.3; 0.003 (2.45 V/m)	van der Meer et al. (2023), Dalecki et al. (2021), Loughran et al. (2019), Hinrikus et al. (2008a, 2009, 2011b, 2017), Roggeveen et al. (2015), Ghosn et al. (2015), Suhhova et al. (2013), Croft et al. (2008, 2010), Bardasano et al. (2007), Regel et al. (2007b), Vecchio et al. (2007), and Krause et al. (2007)
Decreased alpha	2.610 1.947; 1.750	UMTS LTE	<2 1.96; 1.34	Vecsei et al. (2018) and Yang et al. (2017)
Increased beta	1.929–1.939 0.45	WCDMA PM: 7; 14; 21; 40; 70 Hz	0.69; 0.3	Hinrikus et al. (2008b, 2011b, 2017), Roggeveen et al. (2015), Suhhova et al. (2013), and Bachmann et al. (2007)
Decreased beta	2.61	LTE	1.34	Yang et al. (2017)
Increased gamma	1.93–1.94	WCDMA	0.69 0.3	Roggeveen et al. (2015), Curcio et al. (2015), and Hinrikus et al. (2009)
Increased complexity	0.9 0.45	GSM PM	0.5 0.3	Hinrikus et al. (2011a) (HFD), Vecchio et al. (2010) (coherence), and Bachmann et al. (2007) (LDLVP)
No effect	2.45; 1.95; 1.947; 0.38	GSM; UMTS; LTE; WiFi; TETRA	3.75; 2; 1.5; 1; 0.1	Nakatani-Enomoto et al. (2020), Zentai et al. (2015), Eggert et al. (2015), Trunk et al. (2013, 2015), Loughran et al. (2013), and Kleinlogel et al. (2008a)

Most frequently, the increase in alpha power has been evident in resting EEG studies. This can be explained by the structure of the resting-state, eyes closed EEG spectrum, where the alpha power (band 8–12 Hz) is much higher than other bands, and therefore the alterations in alpha power are more easily detectable. Fewer studies reported increased beta band (13–30 Hz) power. A decrease in alpha power was revealed only in two studies and beta power in one study. An increase in gamma band (higher 30 Hz) power was detected in three studies and theta band (4–7 Hz) power in one study. The increase of the EEG signal level dominates in all frequency bands.

TABLE 6.2
Distribution of sleep EEG studies according to the reported radiofrequency radiation effects

Effect	Frequency (GHz)	Modulation	SAR (W/kg)	
Reduced sleep quality; increased gamma, beta, and theta	2.45	WiFi		Bijlsma et al. (2024)
Reduction in alpha during NREM sleep	2.45	WiFi	0.0064	Danker-Hopfe et al. (2020)
No effect on sleep parameters				
Shift in the center frequency of sleep spindles 11–16 Hz toward faster oscillatory activity	3.6; 0.7	5G	<2	Sousouri et al. (2025)
Increased spindles 11–12 Hz	1.93; 1.99	UMTS	1.6	Lowden et al. (2019), Schmid et al. (2012) and Schmid et al. (2012a)
Increased slow-wave activity 0.75–4.5 Hz; reduced motor task	0.9	PM: 0.25–0.8 Hz	0.15	Lustenberger et al. (2013)
Increased delta and theta	0.9	PM: 2 Hz	2	Lustenberger et al. (2015)
Increased delta, theta, alpha	0.9	PM: 2; 8; 14; 217 Hz	2	Schmid et al. (2012b)
No effect	0.08–2.5 (base station); 2.4; 1.95; 1.8; 1.52; 0.96; 0.9	WiFi; GSM; WCDMA PM: 217 Hz;	1; 0, 0.0064	Danker-Hopfe et al. (2010, 2011, 2020), Nakatani-Enomoto et al. (2013), Leitgeb et al. (2008), Regel et al. (2007a), and Fritzer et al. (2007)

These results support the excitation of the brain neuronal oscillations by modulated radiofrequency radiation (Sections 5.8 and 5.9).

Despite the expected nonlinearity of the brain response to radiofrequency radiation, only three studies have used nonlinear EEG analysis methods for the detection of the radiofrequency radiation effect. Bachmann et al. (2007) used a nonlinear method of scaling analysis for the length distribution of low variability periods (LDLVP). The LDLVP method has revealed statistically significant changes for 4 subjects of 15 (26.7%) at 40 Hz modulation frequency. Vecchio et al. (2010) reported that radiofrequency radiation affects inter-hemispheric synchronization of the dominant (alpha) EEG rhythms. Hinrikus et al. (2011b) have successfully used Higuchi's fractal dimension, indicating a statistically significant increase in complexity in all EEG channels with exposure for 12 subjects from 14.

TABLE 6.3
Distribution of cognition, behavior, event-related potential (ERP), and metabolic studies, according to the reported radiofrequency radiation effects

Effect	Frequency (GHz)	Modulation	SAR (W/kg)	Authors
Overnight improvement in the declarative task. Emotional and procedural memory are not affected.	2.45	WiFi	0.025	Bueno-Lopez et al. (2021)
Improved memory and motor tasks	9.25	BS	0.01	Meo et al. (2019)
Visual ERP, increased P1 amplitude, and N1 latency	9.2	GSM	1 and 2	Dalecki et al. (2018)
Reduced reaction time	9.2	GSM	1 and 2	Verrender et al. (2016)
Decrease in correct answers Increased reaction time	0.38	TETRA PM 17.65 Hz	1.5; 6	Sauter et al. (2015)
Pain threshold increases	1.947	UMTS	1.75	Vecsei et al. (2013)
Increased N100 Delayed response	0.894 1.9	GSM WCDMA	0.7 1.7	Leung et al. (2011)
Acoustic ERP, amplitude decrease, adaptation increase	0.9	GSM	0.5	de Tommaso et al. (2009)
Response time change	0.89	GSM	1.09	Luria et al. (2009)
Reducing psychological arousal	0.9	GSM	0.005	Augner et al. (2009)
Attention increase	0.884	GSM	1.09	Wiholm et al. (2009)
Suppresses brain glucose metabolism	0.837	GSM	0.9	Kwon et al. (2011b) and Volkow et al. (2011)
Increased cortex oxygenation	0.9	GSM	1.1	Curcio et al. (2009)

(*Continued*)

TABLE 6.3
(Continued)

Effect	Frequency (GHz)	Modulation	SAR (W/kg)	Authors
No effect	2.45; 2.02; 1.966; 1.8; 1.75; 0.9; 0.88; 0.42	WiFi; GSM; WCDMA; LTE; TETRA BS PM, CW	0.446 W/m^2 10.9; 7.8; 3.75; 2; <2; 1.4; 1.21; 1.18; 1; 0.5	Hosseini et al. (2019), Vecsei et al. (2018), Loughran et al. (2013), Sauter et al. (2011), Riddervold et al. (2010), Kwon et al. (2009, 2010), Eltiti et al. (2009), Cinel et al. (2008), Stefanics et al. (2008), Kleinlogel et al. (2008b), Curcio et al. (2008), Hiller et al. (2008), Regel et al. (2007b), Haarala et al. (2007), and Krause et al. (2007)

Different indicators are used in evaluating sleep EEG and cognition and behavioral effects. Only a few studies have investigated the same parameter. Therefore, the general picture is less consistent than in resting EEG studies.

The physical mechanism of radiofrequency effects on the brain, discussed in Chapter 5, contributes to the interpretation of the effects on the brain neuronal oscillations on the EEG. The mechanisms of the effects behind neuronal oscillations related to sleep, cognition, and behavior are strongly established in the course of long-term development and are little influenced by external factors. Therefore, the radiofrequency radiation effects are more evident in the EEG signal than in cognition and behavior.

This can explain why the alterations in the EEG signal level are dominantly reported in all categories of studies but not the impact on sleep quality or behavior. The effect of radiofrequency radiation that is evident in neuronal activity is most likely compensated and repaired at the level of cognition and human behavior. People cannot be aware of the existence of external inventions. The physical impact of the radiation on neuronal processes is evident but not always transferred to neurophysiological processes of sleep, cognition, and behavior.

The dependences of the effect on the frequency and the level of radiation, as well as on the type of modulation, are diverse. All telecommunication frequency bands and the majority of modulation types and different generations of mobile telecommunication have been reported to cause similar effects. There is no difference between the radiation parameters, frequency, modulation type, and even the specific absorption rate (SAR) between the reported positive and negative results. The lack of regular dependence on radiofrequency can be explained by a relatively small dependence on the origin of the effects, dielectric polarization of tissues, in the radiofrequency band (Chapter 3). The signal structure of different generations of mobile

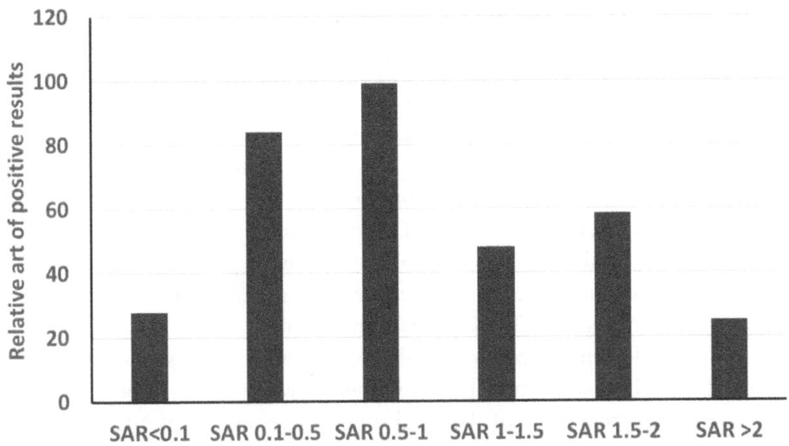

FIGURE 6.2 The rate of positive results at different SAR (W/kg) levels reported in EEG studies, 2007–2024.

telecommunication (Table 6.1) contains combinations of different low frequencies, which impacts is complicated to separate.

A lack of regular dependence on the SAR level is unexpected. The highest SAR values are indicated in studies with non-revealed effects (Table 6.3). The distribution of all EEG studies according to the level of SAR is presented in Figure 6.2. The dependence on SAR is nonmonotonic. Surprisingly, a decreasing trend with increasing levels of radiation is observable.

The inconsistency of applied methods is likely the reason for nonregular dependency on SAR values. Equal SAR values do not describe identical exposure conditions regarding electric field strength because SAR depends on the object under the exposure. SAR describes only the thermal effect. The majority of studies have been performed at low-level radiofrequency radiation SAR < 2, and the thermal impact, if any, is minimal. Dominating at low-level radiation nonthermal effect depends on field strength, not on SAR. The missing indication of field strength or power density in radiofrequency radiation studies is confusing.

The analysis of published results does not allow obtaining exhaustive information about the nature of the effect of radiofrequency radiation because the majority of studies do not report quantitative information about the effect, and the numerical values of the analyzed parameters (e.g. values of changes in the powers of EEG rhythms or other parameters) are missing. Only qualitative information (increase, decrease) is available. The diversity of methods in radiation parameters and the lack of quantitative data about the level of the effect make it difficult to derive conclusions useful for understanding the nature of the effects.

The positive effects reported by several independent studies and research groups support the assertion that radiofrequency radiation affects the human EEG. The effect is more clearly evident in the EEG but is also reported in cognition and behavior.

6.3 ADVANCED METHOD OF DETECTION OF RADIATION-RELATED ALTERATIONS IN EEG

The main limitation in the investigations of radiation effects in the human brain is the high level of chaotic variations of the recorded EEG signals (Figure 4.7) caused by the brain's natural bioelectric activity. These variations, from the point of view of signal processing, can be considered random noise that decreases the possibility of revealing the small alterations in the EEG signal related to the radiation effects. The application of a method discriminating between noise- and effect-related alterations is expected to enable the achievement of more information about the nature of radio-frequency radiation-related effects in the human brain.

The problem of signal reception in the background of noise has been successfully solved in radio astronomy. The radiometer proposed by Robert Henry Dicke (1946) provides the reception of the weak microwave signals at a level much lower than the internal noise on the receiver's input. To reach that, the radiometer Dicke uses low-frequency modulation of a signal on the receiver's input, followed by synchronous detection and integration, to reduce the impact of non-modulated noise (Figure 4.12A). The sensitivity of the radiometer increases as the square root of the integration time.

However, when recording EEG signals, the noise and the signal containing radiation-related alterations originate from the same source, the brain, and cannot be separately treated and separated on the input of the EEG recorder.

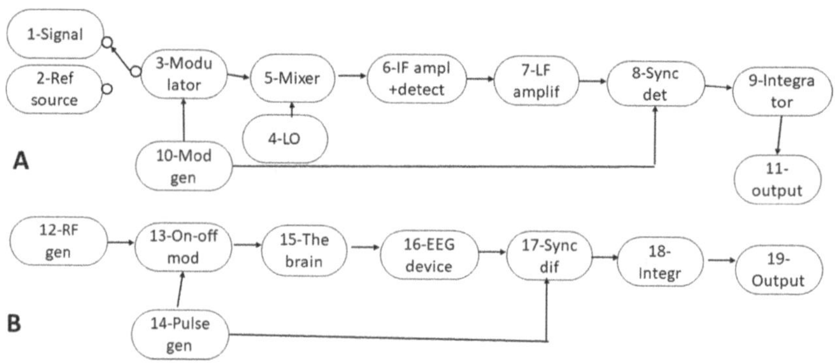

FIGURE 6.3 The basic diagrams of Dicky receiver (A) and the method of integration of differences (B). 1 – the microwave signal from the antenna; 2 – the matched load at room temperature or cooled; 3 – the modulator that switches two inputs at a low modulation frequency; 4 –the mixer that converts the microwave into an intermediate frequency (IF); 5 – local oscillator; 6 – the IF amplifier and the detector that converts the IF to low frequency (LF9); 7 – LF amplifier; 8 – a synchronous detector that distinguishes the modulation frequency signal; 9 – integrator; 10 – generator of modulation frequency; 11 – an output device; 12 – radiofrequency generator; 13 – on-off modulator; 14 – the pulse generator; 15 – the brain; 16 – the EEG device, including electrodes and multichannel amplifiers; 17 – synchronous differentiator between EEG segments corresponding on-off cycles; 18 – integrator summarizing the differences between the on-off cycles; 19 – output device. Steps 13, 17, 18, and 19 are performed on the computer.

A possible solution to the problem is the modulation of the effect directly on the brain using periodic on-off pulse modulation of the applied stimulating radiofrequency radiation (Hinrikus et al., 2007). The modified modulation method (Figure 4.12B) integrates differences in EEG between "on" and "off" radiation periods in a cycle over many cycles. The sensitivity of the method to detect the effect increases as the square root of the number of radiation cycles. The durability of the "on" and "off" cycles of radiation should be longer than the modulation frequency of the radiofrequency radiation and the brain's recovery time. The number of cycles is limited by a reasonable EEG recording time because subjects become drowsy during too-long eyes closed recordings.

The application of the method of integration of differences at fixed parameters of radiation and identical signal analyses can make possible the systematic investigation of the properties of the effect of radiofrequency radiation on the human brain (Chapter 6). The method has been used in the majority of studies by our group. An identical exposure was used in all our studies to make the results comparable. The experiments were performed at a relatively low microwave radiation frequency to increase the depth of penetration. In all our studies, the low-level 450 MHz radiation modulated at 7, 14, 21, 40, 70, 217, or 1000 Hz at the level of power density 1.6 W/m^2 (SAR 0.303 W/kg, field strength 24.5 V/m) was used, except for the study by Suhhova et al. (2013) at two different levels of the radiation. The on-off exposure cycles and sham-exposure sessions were selected by a computer. The computer program also calculated the EEG parameters in the exposure cycles. The experimental protocol was double-blinded.

Reliable knowledge about the properties of radiofrequency radiation effects contributes to the explanation of the properties of the effect of microwave radiation on the brain.

6.4 DEPENDENCE OF THE EFFECT ON THE LEVEL OF MICROWAVE RADIATION

The thermal effect depends on the level of absorbed energy and increases with the level of microwave radiation power. Electric polarization is proportional to electric field strength (Eq. 1.11) and increases with the level of the electric field. The coherent low-level microwave radiation-induced effect is expected to be proportional to the electric field strength. The character of alterations caused by radiofrequency radiation in the EEG at different levels and durations of exposure can provide information about the nature of the effect.

The experimental results with low-level 24.5 V/m 450 MHz radiation modulated at 40 Hz have demonstrated that the distribution of the increase in EEG level caused by the exposure corresponds to the distribution of the radiation power density in the human head (Figure 6.4) (Hinrikus et al., 2011b).

The experiment with low-level 450 MHz radiation modulated at 40 Hz at two different levels of power density, SAR1 = 1.6 W/m^2 (field strength 24.5 V/m) and SAR2 = 0.016 W/m^2 (field strength 2.45 V/m), has been performed by Suhhova et al. (2013). A tenfold reduction in the radiation power density has reduced the level of

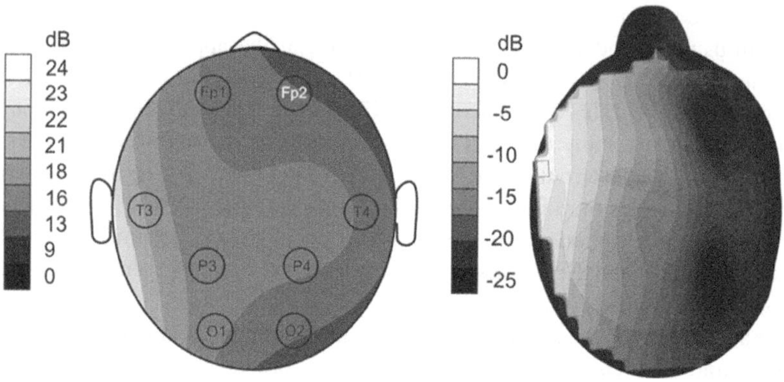

FIGURE 6.4 Left: the distribution of measured relative change in EEG power. Right: the level of modeled radiation power density (from Hinrikus et al., 2011b).

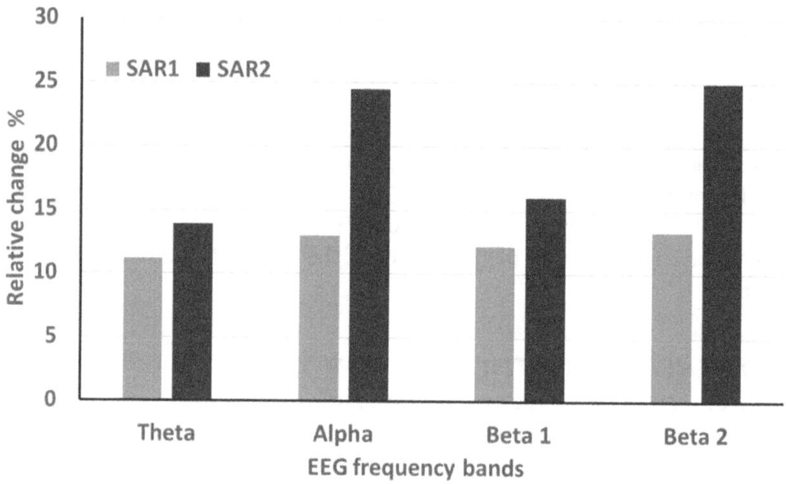

FIGURE 6.5 Relative changes in total EEG between segments with and without exposure at different levels of radiation.

related changes in the EEG and the number of affected subjects but did not exclude the effect. Figure 6.5 presents the result of the study.

The results showed a statistically significant increase in the total EEG at higher radiation levels and a somewhat smaller but still significant increase at lower radiation levels (Figure 6.5). Higher-level radiation causes an approximately five-fold greater increase in EEG levels above sham compared to lower-level radiation. This ratio is less compared to the ratio of electric field strengths and much less than the ratio of power densities between two radiation levels. The decrease in the detected effect compared to the expected one can be caused by the impact of restoration processes. However, the decrease of the effect is much smaller than that related to the thermal effect. An important conclusion is that the effect level follows electric field strength, not power density. This result supports the non-thermal nature of the microwave radiation effect.

Statistically significant changes in EEG have been detected for six individual subjects at the higher radiation level and three subjects at the lower radiation level.

The level of 2.45 W/m^2 is the lowest one at which the microwave radiation effect in the human brain has been detected so far (Hinrikus et al., 2023).

6.5 DEPENDENCE OF THE EFFECT ON MODULATION FREQUENCY

The effect of radiofrequency radiation on the human brain is expected to depend on modulation frequency due to the limited range of physiologic frequencies and the parametric nature of excitation (Section 5.9).

In the studies by Hinrikus et al. (2008a, 2008b) the modulation frequencies have been selected partly within the traditional EEG frequencies (7, 14, 21, and 40 Hz) and partly higher than the traditional EEG frequencies (70, 217, and 1000 Hz). To provide deeper penetration, the relatively low microwave frequency of 450 MHz has been selected. At the used output power of 1W, the maximal measured field power density at the skin was 1.6 W/m^2 and the calculated spatial peak SAR averaged over 1 g was 0.303 W/kg. Ten cycles of the radiation (1 minute on, 1 minute off) have been applied at each of the modulation frequencies. The relative change between the EEG levels at on and off cycles has been selected as an indicator of the effect.

The results of investigations on young healthy volunteers show that the effect decreases at the higher frequencies of modulation (Figure 6.6). The experimental results demonstrate a regular dependence of the radiation-related alteration in the EEG. Despite the alterations in the EEG, none of the subjects declared any feeling related to the exposure. The impact of radiation on the level of the EEG signal has

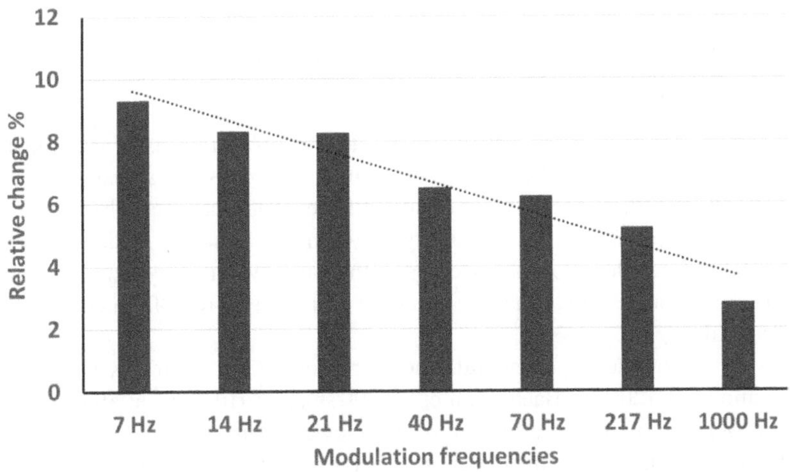

FIGURE 6.6 The effect of the low-level microwave radiation on the EEG signal power at different modulation frequencies.

been strongest at 7, 14, and 21 Hz modulation frequencies located within the spectrum of EEG frequencies, and their contribution to the total EEG power is highest. At these modulation frequencies, the maximum of parametric excitation is expected at 3.5, 7, and 10.5 Hz EEG frequencies (Section 5.9). The frequency 10.5 Hz is located in the alpha band, the maximum of the EEG spectrum, and 21 Hz at the second harmonic of alpha.

A somewhat weaker effect is evident at 40 and 70 Hz modulation frequencies. The strongest excitable frequencies, 20 and 35 Hz, are located higher than the maximum of the EEG spectrum, and therefore less contribution to the neuronal oscillations is expected at these frequencies.

The 217 Hz modulation is expected to stimulate first of all the 108.5 Hz EEG frequency that is even outside the traditional EEG spectrum. However, the alterations in the EEG are still statistically significant. The alteration in the EEG signal related to the 1000 Hz modulation frequency, with the expected maximum at 500 Hz, is not statistically significant. The contribution of single neurons and small neuronal networks, oscillating at frequencies higher than 100 Hz, to the level of the EEG is very small.

The results presented in Figure 6.6 are even somewhat underestimated because the modulation frequencies where the strong impact is expected have been filtered out from the EEG signal to avoid artifacts in the EEG recordings performed during exposure. The modulated radiation is expected to generate strong excitation in the EEG at the modulation frequency (Section 5.9). These alterations in the EEG have contributed only partly to the results due to the wider zone of excitation around the filtered-out excitation frequency.

The results of the experimental studies demonstrate that the effect of modulated radiofrequency radiation has a regular dependence on the modulation frequency and is stronger at the modulation frequencies within the spectrum of natural physiological frequencies.

6.6 INDIVIDUAL SENSITIVITY TO LOW-LEVEL MICROWAVE RADIATION

Each person is individual and differs from other people in terms of their anatomy, physiology, and behavior. The reaction of the human organism to an external stressor is a complex, nonlinear process comprising different subprocesses, which in turn are complex and nonlinear. A big variability between individuals is evident. The probability that all people react to a stressor in the same way is very small. Therefore, there is no reason to expect that the reactions of people to radiofrequency radiation are identical.

The results of the experimental studies performed by Hinrikus et al. (2004), Bachmann et al. (2007a), Bachmann et al (2007b), and Hinrikus et al. (2008) show different sensitivity of subjects to radiofrequency exposure.

The results demonstrate big differences between the responses to low-level radiofrequency radiation by individual subjects (Figure 6.7). The response is somewhat different at different modulation frequencies (Hinrikus et al., 2008b). However, the same subjects are affected at different modulation frequencies.

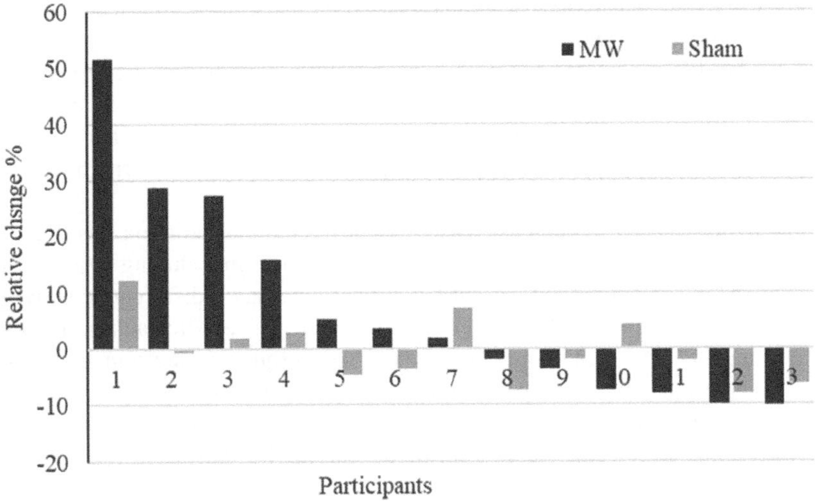

FIGURE 6.7 Relative alterations regarding sham exposure in the total EEG power of individual subjects at radiofrequency exposure (450 MHz, 1.6 W/m^2, modulation 14 Hz).

TABLE 6.4
The number of subjects affected by radiofrequency radiation (450 MHz, 1.6W/m2) at different modulation frequencies (data from Hinrikus et al., 2008b)

Modulation frequency (Hz)	Number of subjects	Number of affected subjects	Rate of affected subjects (%)
14	13	4	31
21	13	3	23
40	15	3	20
70	15	2	13
217	19	3	16
1000	19	0	0

The numbers in Table 6.4 show that the rate of affected subjects varies between modulation frequencies.

The number of subjects affected by radiofrequency exposure decreases at higher modulation frequencies. The rate is higher at modulation frequencies within the spectrum of physiological frequencies 14–40 Hz, decreases at frequencies higher than the traditional EEG spectrum 70–217 Hz, and is zero at 1000 Hz modulation. As expected, the rate of affected subjects is correlated with the sensitivity to modulation frequency discussed in Section 6.5. The rate of those affected by radiofrequency radiation among young healthy people, estimated to be between 13% and 31%, is even higher compared to the rate of multiple chemical sensitivity, estimated to be between 2% and 10% in the general population.

The discussed sensitivity is estimated based on the short-term effect that diminishes very quickly, as shown in Section 6.8. The human response to long-term exposure can be different. Unfortunately, only very little information is available about the effect of long-term microwave radiation on humans. Most probably, the reparation mechanisms of some people can cope with long-term exposure, but some other people cannot. The people's ability to cope with the impact of external factors like radiation, chemicals, or viruses is individual.

The hypothesis of hypersensitivity to electromagnetic radiation has been proposed by a number of researchers. There is no point in introducing hypersensitivity as a special phenomenon. It's just that everyone has a different sensitivity to microwave radiation, just like other stressors and viruses. Nowadays, all people are permanently living in stronger or weaker radiofrequency electromagnetic fields of various frequencies, from power electricity to microwaves. The differentiation between the disorders related to the radiation or some other factors would be possible using large-scale population investigations but hardly at the level of an individual.

6.7 EFFECT ON ATTENTION AND MEMORY IN VISUAL PERCEPTION

The ability of radiofrequency radiation to generate disturbances in EEG oscillations can be considered as a noise regarding regular brain functioning. The additional noise in the brain affects the perception and cognitive processes in the brain (Table 5.3). The rate of positive results in microwave effects on behavior and cognition is lower than that on resting EEG (Figure 6.1). The results of the study by Lass et al. (2002) show that the reported effect varies in cognitive tasks of different complexity. The effect may cause a decrease in the number of errors for less complicated neuropsychological tasks and an increase in errors for more complicated tasks. To reveal the veiled impacts of microwave radiation on the neurophysiological processes, the more specific and complicated methods would benefit.

A new method to study the possible effects on central nervous system functioning, face masking, is presented (Bachmann, 1994). The psychophysiological phenomenon named visual masking is a cognitive task of a high level of complexity. Face masking is observable when two or more extremely short-term visual stimulus images interact within a short time interval. Visual masking can be defined as the impairment or total suppression of the perception of a transient visual stimulus/image A as a consequence of the exposure of another visual stimulus/image B immediately after A.

The experiment was performed by Rodina et al. (2005) on ten healthy young volunteers using low-level 7 Hz-modulated, 450 MHz radiofrequency radiation at a power density of 1.6 W/m^2. The subjects were presented with two successive photos of a male face, and the task of the subject was to identify both pictures from a group of six alternative photos in the right order. A subject made 800 trials, 400 with and 400 without microwave exposure. The recognition of both stimuli in the correct order has been reported to be higher in sham exposure compared to radiofrequency exposure. The difference was only 5% but statistically significant. There was no statistical significance between sham and microwave results in partly correct answers. It can be concluded that the early stages of visual information processing are overwhelmingly robust and routine activities, so the low-level microwave radiation upon it is weak. The effect becomes evident not in perception but in the next step of neurophysiological processes.

The results of the study by Rodina et al. demonstrate that the effect on visual perception, attention, and memory is weaker compared to the effect on resting EEG reported by the same group (Hinrikus et al., 2008a). This conclusion is consistent with the trend mentioned in Table 6.3. The possible reason is the nature of neurophysiologic processes that have developed and established over a long period of development, and the corresponding neural connections are strong. The radiation can affect rather later stages of information processing and becomes evident in the case of the most complicated tasks. A weak external stressor can affect resting-state EEG and stimulate neuronal oscillations, but the probability of breaking the established strong neurophysiologic connections is very low.

6.8 RECOVERY AND AFTEREFFECT

The human organism can repair the alterations caused by external disturbing factors and restore the normal state. The ability of the brain to restore the normal level of the EEG signal under exposure to microwave radiation has become evident in the study by Hinrikus et al. (2008a), where the statistically significant effect was detected only for the first half-period, 0–30 seconds, of 1 minute excitation cycles. For the second half-period, 30–60 seconds, of the excitation cycles, the effect has been much less and statistically insignificant. The relative alterations in the total EEG power between cycles 1 minute with and 1 minute without radiation presented

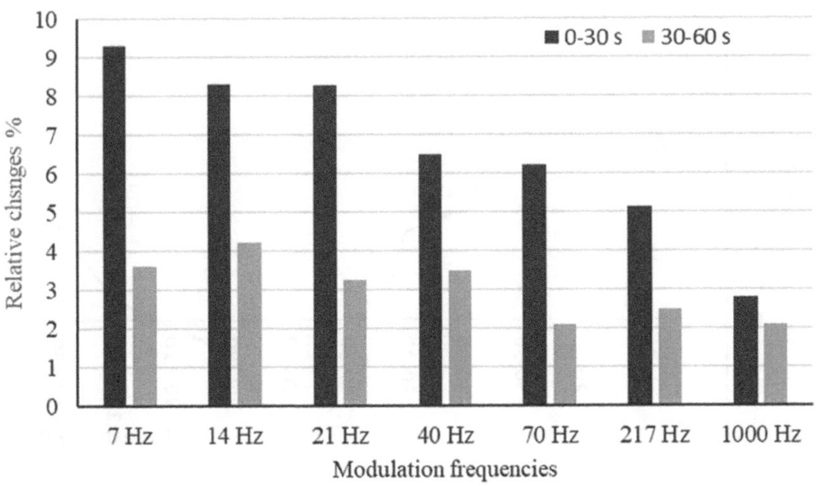

FIGURE 6.8 The effect of the low-level radiofrequency radiation on the EEG signal power during the first 0–30 seconds and second 30–60 seconds half-periods of the 1 minute radiation cycle at different modulation frequencies.

in Figure 6.7 differ drastically in the case of the first (0–30 seconds) and second (30–60 seconds) half-period of the radiation cycles. The effect that has been statistically significant at all modulation frequencies (except 1000 Hz) during the first 30 seconds of the cycles (Figure 6.4) is much less during the second half-period of radiation.

The impact of the recovery process, due to its inertia, can be evident after the end of the exposure. The detailed analysis of the temporal dynamics of the radiofrequency radiation effect during the cycles with and without radiation makes visible an aftereffect, the decrease of the EEG level after the exposure (Figure 6.8).

The graft in Figure 6.9 demonstrates the nonmonotonic time course of the excitation as well as recovery processes. The excitation reaches the maximal level during about a 20-second delay from the onset of radiation. After that, the excitation level lowers most probably due to the recovery mechanisms. The recovery mechanisms compensate and even overcompensate for the impact of radiation. The EEG level lowers compared to the reference one and reaches the minimal level after about a 20-second interval after radiation shutdown. The 20-second delay in EEG excitation and recovery maximums is determined by the time constant of the processes related to the impact of microwave radiation on bioelectrical oscillations in the human brain. Most interesting in the time course of excitation is the decrease of EEG level after the end of stimulation, lowering its normal undisturbed level – aftereffect.

The aftereffect is known to appear in ERPs in visual perception. The study by Bachmann et al. (2018), applying the ERP technique, has demonstrated that aftereffects also become evident using pulses of microwave radiation as a "visual" stimulus. The EEG responses locked to radiation stimulus averaged over about 130 stimuli have detected the enhanced EEG signal level and about twice the signal-to-noise

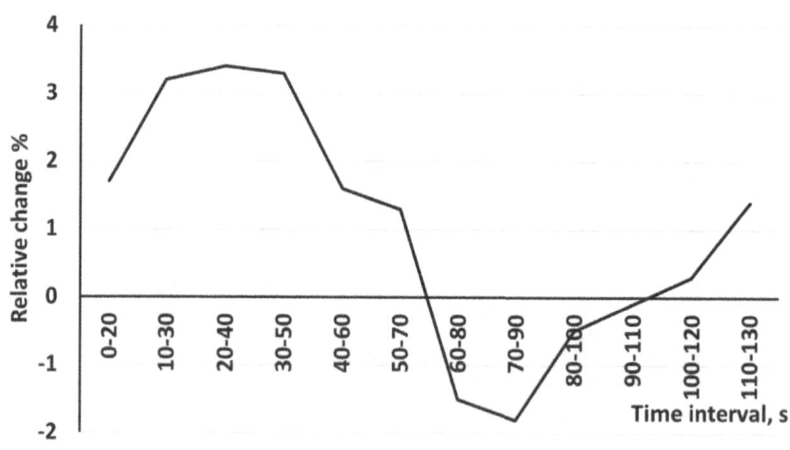

FIGURE 6.9 The relative difference in the total EEG regarding the reference level in 10 seconds shifted overlapping 20-second windows during a cycle (60 seconds on and 60 seconds off) averaged over 10 cycles.

ratio at 200–440 ms after the stimulus. The result is consistent with the aftereffect caused by a visual stimulus. The evidence of aftereffects related to microwave radiation supports the possibility of microwave radiation-induced ERP in EEG signals.

6.9 CONCLUSIONS

The series of investigations on low-level microwave effects on the human brain performed by the author's research group using a special method for the suppression of the impact of natural EEG fluctuations made it possible to determine some characteristic features of the effect.

- The low-level microwave effect is related to electric field strength, not the power of the radiation, because dielectric polarization is related to the electric field strength.
- The microwave effect does not depend much on carrier frequency because dielectric polarizations vary slowly within the radiofrequency band. The dependence on modulation frequency is important, and the effect is stronger at the frequencies within the spectrum of physiological frequencies.
- The sensitivity to microwave radiation is different in various people. The impact of identical microwave radiation causes very different alterations in the EEG of healthy individuals. There is no point in introducing hypersensitivity as a special phenomenon. Everyone has a different sensitivity to microwave radiation like other stressors and viruses.
- The impact of low-level microwave radiation can become evident in complicated neurophysiological tasks. Early stages of perception have developed and established over a long period of development, and the corresponding neural connections are strong. The later stages related to the processing of nontraditional information in the brain can be more sensitive to external stressors.
- The recovery mechanism compensates for the changes related to low-level microwave radiation in the EEG signal. Due to the inertia of the recovery processes, the aftereffect appears in the EEG signal, which makes the impact of microwave pulses similar to that of visual stimuli.

REFERENCES

Augner C, Florian M, Pauser G, Oberfeld G, Hacker GW, (2009). GSM base stations: short-term effects on well-being. *Bioelectromagnetics*. 30(1):73–80.

Bachmann M, Hinrikus H, Aadamsoo K, Vohma U, Lass J, Rubljova J, Suhhova A, Tuulik V, (2007b). Modulated microwave effects on individuals with depressive disorder. *Environmentalist*. 27(4):505–510.

Bachmann M, Rubljova J, Lass J, Tomson R, Tuulik V, Hinrikus H, (2007). Adaptation of human brain bioelectrical activity to low-level microwave. *Annu Int Conf IEEE Eng Med Biol Soc*. 4747–4750.

Bachmann M, Tomson R, Kalda J, Säkki M, Lass J, Tuulik V, Hinrikus H, (2007a). Individual changes in human EEG caused by 450 MHz microwave modulated at 40 and 70 Hz. *Environmentalist*. 27(4):511–517.

Bachmann M, Päeske L, Ioannides AA, Lass J, Hinrikus H, (2018). After-effect induced by microwave radiation in human electroencephalographic signal: a feasibility study. *Int J Radiat Biol.* 94:896–901.

Bachmann T, (1994). *Psychophysiology of visual masking.* Nova Science Publishers, New York.

Bardasano JL, Alvarez-Ude J, Gutierrez I, Raposo M, Goya R, (2007). EEG bioeffects on cochlear deaf from cellular phones. *Environmentalist.* 27(4):519–523.

Bijlsma N, Conduit R, Kennedy G, Cohen M, (2024). Does radiofrequency radiation impact sleep? A double-blind, randomised, placebo-controlled, crossover pilot study. *Front Public Health.* 12:1481537. doi: 10.3389/fpubh.2024.1481537

Bueno-Lopez A, Eggert T, Dorn H, Schmid G, Hirtl R, Danker-Hopfe H, (2021). Effects of 2.45 GHz Wi-Fi exposure on sleep-dependent memory consolidation. *J Sleep Res.* 30:e13224. doi: 10.1111/jsr.13224.

Cinel C, Boldini A, Fox E, Russo R, (2008). Does the use of mobile phones affect human short-term memory or attention? *Appl Cognit Psychol.* 22(8):1113–1125.

Croft RJ, Hamblin DL, Spong J, Wood AW, McKenzie RJ, Stough C, (2008). The effect of mobile phone electromagnetic fields on the alpha rhythm of human electroencephalogram. *Bioelectromagnetics.* 29(1):1–10.

Croft RJ, Leung S, McKenzie RJ, Loughran SP, Iskra S, Hamblin DL, Cooper NR, (2010). Effects of 2G and 3G mobile phones on human alpha rhythms: resting EEG in adolescents, young adults, and the elderly. *Bioelectromagnetics.* 31(6):434–444.

Curcio G, Valentini E, Moroni F, Ferrara M, De Gennaro L, Bertini M, (2008). Psychomotor performance is not influenced by brief repeated exposures to mobile phones. *Bioelectromagnetics.* 29(3):237–241.

Curcio G, Ferrara M, Limongi T, Tempesta D, Di Sante G, De Gennaro L, Quaresima V, Ferrari M, (2009). Acute mobile phones exposure affects frontal cortex hemodynamics as evidenced by functional near-infrared spectroscopy. *J Cereb Blood Flow Metab.* 29(5):903–910.

Curcio G, Mazzucchi E, Della Marca G, Vollono C, Rossini PM, (2015). Electromagnetic fields and EEG spiking rate in patients with focal epilepsy. *Clin Neurophysiol.* 126(4): 659–666.

Dalecki A, Loughran SP, Verrender A, Burdon CA, Taylor NAS, Croft RJ, (2018). Does acute radio-frequency electromagnetic field exposure affect visual event-related potentials in healthy adults? *Clin Neurophysiol.* 129(5):901–908.

Dalecki A, Verrender A, Loughran SP, Croft RJ, (2021). The effect of GSM electromagnetic field exposure on the waking electroencephalogram: methodological influences. *Bioelectromagnetics.* May;42(4):317–328.

Danker-Hopfe H, Dorn H, Bornkessel C, Sauter C, (2010). Do mobile phone base stations affect sleep of residents? Results from an experimental double-blind sham-controlled field study. *Am J Hum Biol.* 22(5):613–618.

Danker-Hopfe H, Dorn H, Bahr A, Anderer P, Sauter C, (2011). Effects of electromagnetic fields emitted by mobile phones (GSM 900 and WCDMA/UMTS) on the macrostructure of sleep. *J Sleep Res.* 20(1):73–81.

Danker-Hopfe H, Bueno-Lopez A, Dorn H, Schmid G, Hirtl R, Eggert T, (2020). Spending the night next to a router - Results from the first human experimental study investigating the impact of Wi-Fi exposure on sleep. *Int J Hyg Environ Health.* July;228:113550.

de Tommaso M, Rossi P, Falsaperla R, Francesco Vde V, Santoro R, Federici A, 2009. Mobile phones exposure induces changes of contingent negative variation in humans. *Neurosci Lett.* 464(2):79–83.

Dicke, R. H, (1946). The measurement of thermal radiation at microwave frequencies. *Rev Sci Inst.* 17:268–275.

Eggert T, Dorn H, Sauter C, Marasanov A, Hansen ML, Peter A, Schmid G, Bolz T, Danker-Hopfe H, (2015). Terrestrial Trunked Radio (TETRA) exposure and its impact on slow cortical potentials. *Environ Res.* 143 Pt A:112–122.

Eltiti S, Wallace D, Ridgewell A, Zougkou K, Russo R, Sepulveda F, Fox E, (2009). Short-term exposure to mobile phone base station signals does not affect cognitive functioning or physiological measures in individuals who report sensitivity to electromagnetic fields and controls. *Bioelectromagnetics.* 30(7):556–563.

Fritzer G, Göder R, Friege L, Wachter J, Hansen V, Hinze-Selch D, Aldenhoff JB, (2007). Effects of short- and long-term pulsed radiofrequency electromagnetic fields on night sleep and cognitive functions in healthy subjects. *Bioelectromagnetics.* 28(4):316–325.

Ghosn R, Yahia-Cherif L, Hugueville L, Ducorps A, Lemarechal JD, Thuroczy G, de Seze R, Selmaoui B, (2015). Radiofrequency signal affects alpha band in resting electroencephalogram. *J Neurophysiol.* 113(7): 2753–2759.

Gutiérrez JS, Rodríguez de Fonseca F, Rubio G, (2016).Cell-phone addiction: a review. *Front Psychiatry.* 7:175. doi: 10.3389/fpsyt.2016.00175.

Haarala C, Takio F, Rintee T, Laine M, Koivisto M, Revonsuo A, Hämäläinen H, (2007). Pulsed and continuous wave mobile phone exposure over left versus right hemisphere: effects on human cognitive function. *Bioelectromagnetics.* 28(4):289–295.

Hasted JB, (1973). *Aqueous dielectrics.* Chapman and Hall, London, 302p.

Hillert L, Akerstedt T, Lowden A, Wiholm C, Kuster N, Ebert S, Boutry C, Moffat SD, Berg M, Arnetz BB, (2008). The effects of 884 MHz GSM wireless communication signals on headache and other symptoms: an experimental provocation study. *Bioelectromagnetics.* 29(3):185–196.

Hinrikus H, Parts M, Lass J, Tuulik V, (2004). Changes in human EEG caused by low-level modulated microwave stimulation. *Bioelectromagnetics.* 25(6):431–440.

Hinrikus H, Bachmann M, Kalda J, Sakki M, Lass J, Tomson R, (2007). Methods of electroencephalographic signal analysis for detection of small hidden changes. *Nonlinear Biomed Phys.* 1(1):9. doi: 10.1186/1753–4631-1–9.

Hinrikus H, Bachmann M, Lass J, Tomson R, Tuulik V, (2008a). Effect of 7, 14 and 21 Hz modulated 450 MHz microwave radiation on human electroencephalographic rhythms. *Int J Radiat Biol.* 84(1):69–79.

Hinrikus H, Bachmann M, Lass J, Karai D, Tuulik V, (2008b). Effect of low frequency modulated microwave exposure on human EEG: Individual sensitivity. *Bioelectromagnetics.* 29(7):527–538.

Hinrikus H, Bachmann M, Lass J, Tuulik V, (2009). Effect of modulated at different low frequencies microwave radiation on human EEG. *Environmentalist.* 29(2):215–219.

Hinrikus H, Bachmann M, Lass J, (2011a). Parametric mechanism of excitation of the electroencephalographic rhythms by modulated microwave radiation. *Int J Radiat Biol.* 87(11):1077–1085.

Hinrikus H, Bachmann M, Karai D, Klonowski W, Lass J, Stepien P, Stepien R, Tuulik V, (2011b). Higuchi's fractal dimension for analysis of the effect of external periodic stressor on electrical oscillations in the brain. *Med Biol Eng Comput.* 49(5):585–591.

Hinrikus H, Bachmann M, Karai D, Lass J. (2017). Mechanism of low-level microwave radiation effect on nervous system. *Electromagn Biol Med.* 36(2):202–212.

Hinrikus H, Lass J, Bachmann M, (2021). Threshold of radiofrequency electromagnetic field effect on human brain. *Int J Radiat Biol.* 97(11):1505–1515.

Hinrikus H, Koppel T, Lass J, Orru H, Roosipuu P, Bachmann M, (2022). Possible health effects on the human brain by various generations of mobile telecommunication: a review based estimation of 5G impact. *Int J Radiat Biol.* 98(7):1210–1221.

Hinrikus H, Koppel T, Lass J, Roosipuu P, Bachmann M, (2023). Limiting exposure to radiofrequency radiation: the principles and possible criteria for health protection. *Int J Radiat Biol.* 99(8):1167–1177.

Kleinlogel H, Dierks T, Koenig T, Lehmann H, Minder A, Berz R, (2008a). Effects of weak mobile phone - electromagnetic fields (GSM, UMTS) on well-being and resting EEG. *Bioelectromagnetics*. 29(6):479–487.

Kleinlogel H, Dierks T, Koenig T, Lehmann H, Minder A, Berz R, (2008b). Effects of weak mobile phone - electromagnetic fields (GSM, UMTS) on event related potentials and cognitive functions. *Bioelectromagnetics*. 29(6):488–497.

Krause CM, Pesonen M, Haarala Björnberg C, Hämäläinen H, (2007). Effects of pulsed and continuous wave 902 MHz mobile phone exposure on brain oscillatory activity during cognitive processing. *Bioelectromagnetics*. 28(4):296–230.

Kwon MS, Kujala T, Huotilainen M, Shestakova A, Naatanen R, Hämäläinen H, (2009). Preattentive auditory information processing under exposure to the 902 MHz GSM mobile phone electromagnetic field: a mismatch negativity (MMN) study. *Bioelectromagnetics*. 30(3):241–248.

Kwon MS, Huotilainen M, Shestakova A, Kujala T, Näätänen R, Hämäläinen H, (2010). No effects of mobile phone use on cortical auditory change-detection in children: an ERP study. *Bioelectromagnetics*. 31(3):191–199.

Kwon MS, Hämäläinen H, (2011a). Effects of mobile phone electromagnetic fields: critical evaluation of behavioral and neurophysiological studies. *Bioelectromagnetics*. 32:253–272.

Kwon MS, Vorobyev V, Kännälä S, Laine M, Rinne JO, Toivonen T, Johansson J, Teras M, Lindholm H, Alanko T, Hamalainen H, (2011b). GSM mobile phone radiation suppresses brain glucose metabolism. *J Cereb Blood Flow Metab*. 31(12):2293–2301.

Lai H, Levitt BB, (2022). The roles of intensity, exposure duration, and modulation on the biological effects of radiofrequency radiation and exposure guidelines. *Electromagn Biol Med*. 41 (2):230–255.

Lass J, Tuulik V, Ferenets R, Riisalo R, Hinrikus H, (2002). Effects of 7 Hz-modulated 450 MHz electromagnetic radiation on human performance in visual memory tasks. *Int J Radiat Biol*. 78(10):937–944.

Leitgeb N, Schröttner J, Cech R, Kerbl R, (2008). EMF-protection sleep study near mobile phone base stations]. *Somnologie (Somnology)*. 12(3):234–243.

Leung S, Croft RJ, McKenzie RJ, Iskra S, Silber B, Cooper NR, O'Neill B, Cropley V, Diaz-Trujillo A, Hamblin D, Simpson D, (2011). Effects of 2G and 3G mobile phones on performance and electrophysiology in adolescents, young adults and older adults. *Clin Neurophysiol*. 122(11):2203–2216.

Loughran SP, McKenzie RJ, Jackson ML, Howard ME, Croft RJ, (2012). Individual differences in the effects of mobile phone exposure on human sleep: rethinking the problem. *Bioelectromagnetics*. 33(1):86–93.

Loughran SP, Benz DC, Schmid MR, Murbach M, Kuster N, Achermann P, (2013). No increased sensitivity in brain activity of adolescents exposed to mobile phone-like emissions. *Clin Neurophysiol*. 124(7):1303–1308.

Loughran SP, Verrender A, Dalecki A, Burdon CA, Tagami K, Park J, Taylor NAS, Croft RJ, (2019). Radiofrequency electromagnetic field exposure and the resting EEG: exploring the thermal mechanism hypothesis. *Int J Environ Res Public Health*. 16(9):e1505.

Lowden A, Nagai R, Åkerstedt T, Hansson Mild K, Hillert L, (2019). Effects of evening exposure to electromagnetic fields emitted by 3G mobile phones on health and night sleep EEG architecture. *J Sleep Res*. 28(4):e12813.

Luria R, Eliyahu I, Hareuveny R, Margaliot M, Meiran N, (2009). Cognitive effects of radiation emitted by cellular phones: the influence of exposure side and time. *Bioelectromagnetics*. 30(3):198–204.

Lustenberger C, Murbach M, Dürr R, Schmid MR, Kuster N, Achermann P, Huber R, (2013). Stimulation of the brain with radiofrequency electromagnetic field pulses affects sleep-dependent performance improvement. *Brain Stimul*. 6(5):805–811.

Lustenberger C, Murbach M, Tushaus L, Wehrle F, Kuster N, Achermann P, Huber R, (2015). Inter-individual and intra-individual variation of the effects of pulsed RF EMF exposure on the human sleep EEG. *Bioelectromagnetics*. 36(3):169–177.

Meo SA, Almahmoud M, Alsultan Q, Alotaibi N, Alnajashi I, Hajjar WM, (2019). Mobile phone base station tower settings adjacent to school buildings: impact on students' cognitive health. *Am J Mens Health*. 13(1):1557988318816914.

Nakatani-Enomoto S, Furubayashi T, Ushiyama A, Groiss SJ, Ueshima K, Sokejima S, Simba AY, Wake K, Watanabe SI, Nishikawa M, Miyawaki K, Taki M, Ugawa Y, (2013). Effects of electromagnetic fields emitted from W-CDMA-like mobile phones on sleep in humans. *Bioelectromagnetics*. 34(8):589–598.

Nakatani-Enomoto S, Yamazaki M, Nishiura K, Enomoto H, Ugawa Y, (2020). Effects of electromagnetic fields from long-term evolution on awake electroencephalogram in healthy humans. *Neurosci Res*. 156:102–107.

Regel SJ, Gottselig JM, Schuderer J, Tinguely G, Retey JV, Kuster N, Landolt HP, Achermann P, (2007a). Pulsed radio frequency radiation affects cognitive performance and the waking electroencephalogram. *NeuroReport*. 18:803–807.

Regel SJ, Tinguely G, Schuderer J, Adam M, Kuster N, Landolt HP, Achermann P, (2007b). Pulsed radio-frequency electromagnetic fields: dose-dependent effects on sleep, the sleep EEG and cognitive performance. *J Sleep Res*. 16:253–258.

Riddervold IS, Kjaergaard SK, Pedersen GF, Andersen NT, Franek O, Pedersen AD, Sigsgaard T, Zachariae R, Molhave L, Andersen JB, (2010). No effect of TETRA hand portable transmission signals on human cognitive function and symptoms. *Bioelectromagnetics*. 31(3):191–199.

Rodina A, Lass J, Riipulk J, Bachmann T, Hinrikus H, (2005). Study of effects of low level microwave field by method of face masking. *Bioelectromagnetics*. 26(7):571–577.

Roggeveen S, van Os J, Viechtbauer W, Lousberg R, (2015). EEG changes due to experimentally induced 3G mobile phone radiation. *PLoS One*. 10(6):e0129496.

Sauter C, Dorn H, Bahr A, Hansen ML, Peter A, Bajbouj M, Danker-Hopfe H, (2011). Effects of exposure to electromagnetic fields emitted by GSM 900 and WCDMA mobile phones on cognitive function in young male subjects. *Bioelectromagnetics*. 32(3): 179–190.

Sauter C, Eggert T, Dorn H, Schmid G, Bolz T, Marasanov A, Hansen ML, Peter A, Danker-Hopfe H, (2015). Do signals of a hand-held TETRA transmitter affect cognitive performance, well-being, mood or somatic complaints in healthy young men? Results of a randomized double-blind cross-over provocation study. *Environ Res*. 140:85–94.

Schmid MR, Loughran SP, Regel SJ, Murbach M, Bratic Grunauer A, Rusterholz T, Bersagliere A, Kuster N, Achermann P, (2012). Sleep EEG alterations: effects of different pulse-modulated radio frequency electromagnetic fields. *J Sleep Res*. 21(1):50–58.

Schmid MR, Murbach M, Lustenberger C, Maire M, Kuster N, Achermann P, Loughran SP, (2012a). Sleep EEG alterations: effects of pulsed magnetic fields versus pulse-modulated radio frequency electromagnetic fields. *J Sleep Res*. 21(6):620–629.

Schmid MR, Loughran SP, Regel SJ, Murbach M, Bratic Grunauer A, Rusterholz T, Bersagliere A, Kuster N, Achermann P, (2012b). Sleep EEG alterations: effects of different pulse-modulated radio frequency electromagnetic fields. *J Sleep Res*. 21(1):50–58.

Sousouri G, Eicher C, MariaAngelo RD, Billecocq M, Fussinger T, Studler M, Capstick M, Kuster N, Achermann P, Huber R, Landolt H, (2025). 5G radio-frequency-electromagnetic-field effects on the human sleep electroencephalogram: a randomized controlled study in CACNA1C genotyped volunteers. *NeuroImage*, 121340. doi: 10.1016/j.neuroimage.2025.121340

Stefanics G, Thuroczy G, Kellenyi L, Hernadi I, (2008). Effects of twenty-minute 3G mobile phone irradiation on event related potential components and early gamma synchronization in auditory oddball paradigm. *Neuroscience*. 157(2):453–462.

Suhhova A, Bachmann M, Karai D, Lass J, Hinrikus H, (2013). Effect of microwave radiation on human EEG at two different levels of exposure. *Bioelectromagnetics*. 34(4):264–274.

Trunk A, Stefanics G, Zentai N, Kovacs-Balint Z, Thuroczy G, Hernadi I, (2013). No effects of a single 3G UMTS mobile phone exposure on spontaneous EEG activity, ERP correlates, and automatic deviance detection. *Bioelectromagnetics*. 34(1):31–42.

Trunk A, Stefanics G, Zentai N, Bacskay I, Felinger A, Thuroczy G, Hernadi I, (2015). Effects of concurrent caffeine and mobile phone exposure on local target probability processing in the human brain. *Sci Rep*. 5:14434.

Valentini E, Curcio G, Moroni F, Ferrara M, De Gennaro L, Bertini M, (2007). Neurophysiological effects of mobile phone electromagnetic fields on humans: a comprehensive review. *Bioelectromagnetics*. 28:415–432.

van der Meer JN, Eisma YB, Meester R, Jacobs M, Nederveen AJ, (2023). Effects of mobile phone electromagnetic fields on brain waves in healthy volunteers. *Sci Rep*. 13(1):21758. doi: 10.1038/s41598-023-48561-z.

Vecchio F, Babiloni C, Ferreri F, Curcio G, Fini R, Del Percio C, Rossini PM, (2007). Mobile phone emission modulates interhemispheric functional coupling of EEG alpha rhythms. *Eur J Neurosci*. 25(6):1908–1913.

Vecchio F, Babiloni C, Ferreri F, Buffo P, Cibelli G, Curcio G, van Dijkman S, Melgari JM, Giambattistelli F, Rossini PM, (2010). Mobile phone emission modulates interhemispheric functional coupling of EEG alpha rhythms in elderly compared to young subjects. *Clin Neurophysiol*. 121(2):163–171.

Vecsei Z, Csatho A, Thuroczy G, Hernadi I, (2013). Effect of a single 30 min UMTS mobile phone-like exposure on the thermal pain threshold of young healthy volunteers. 2013. *Bioelectromagnetics*. 34(7):530–541.

Vecsei Z, Knakker B, Juhász P, Thuróczy G, Trunk A, Hernádi I, (2018). Short-term radiofrequency exposure from new generation mobile phones reduces EEG alpha power with no effects on cognitive performance. *Sci Rep*. 8(1):18010.

Verrender A, Loughran SP, Dalecki A, McKenzie R, Croft RJ, (2016). Pulse modulated radiofrequency exposure influences cognitive performance. *Int J Radiat Biol*. 92(10):603–610.

Volkow ND, Tomasi D, Wang GJ, Vaska P, Fowler JS, Telang F, Alexoff D, Logan J, Wong C, (2011). Effects of cell phone radiofrequency signal exposure on brain glucose metabolism. *JAMA*. 305(8):808–813.

Wiholm C, Lowden A, Kuster N, Hillert L, Arnetz BB, Akerstedt T, Moffat SD, (2009). Mobile phone exposure and spatial memory. *Bioelectromagnetics*. 30(1):59–65.

Yang L, Chen Q, Lv B, Wu T, (2017). Long-Term Evolution electromagnetic fields exposure modulates the resting state EEG on alpha and beta bands. *Clin EEG Neurosci*. 48(3):168–175.

Zentai N, Csatho A, Trunk A, Fiocchi S, Parazzini M, Ravazzani P, Thuroczy G, Hernadi I, (2015). No effects of acute exposure to Wi-Fi electromagnetic fields on spontaneous EEG activity and psychomotor vigilance in healthy human volunteers. *Radiat Res*. 184(6):568–577.

Part III

Biological and health effects of environmental microwave radiation and health protection limits

The displacement of a single electron by a billionth of a centimeter at one moment might make the difference between a man being killed by an avalanche a year later, or escaping.

Alan Turing

DOI: 10.1201/9781003729976-9

7 Biological and health effects

Assessment of certainty, carcinogenicity, and medical application

HIGHLIGHTS

- The majority of the studies on different biological effects of low-level microwave radiation in various cells and tissues by hundreds of researchers have reported positive results.
- The contradictory and heterogeneous results of health effect investigations reflect the nonlinearity of living systems.
- The transfer of biological effects to the impact on health is a chaotic, nonlinear process with unpredictable outcomes.
- A long latency period makes it impossible to determine the initial change which has triggered the disease.
- The impact of low-level microwave radiation on health is still under discussion.
- The selection of studies for the analysis of exposure to modulated or continuous-wave radiation can result in controversial conclusions on biological effects in reviews.

7.1 INTRODUCTION

The potential impact of low-level microwave radiation on human health is a significant concern and a matter of public interest. Recent books by Barnes and Greenebaum (2018) and Panagopoulos (2023) present exhaustive reviews on various aspects of radiofrequency radiation health effects. The current chapter aims not to review the huge amount of data from the performed studies but to analyze the results of the reviews published on various topics of radiofrequency radiation effects to discuss the health effects, considering the basic knowledge presented in the first part of this book and the physical model in Section 5.13.

Dielectric polarization is the origin of the physical model of the microwave radiation effect (Section 5.13). A logical series of processes in tissues initiated by dielectric polarization can lead to physiological and biological changes and finally health effects (Figure 5.7). The alterations in hydrogen bonds caused by dielectric

 DOI: 10.1201/9781003729976-10

polarization can result in a wide spectrum of possible physiological and biological effects. Therefore, it is not surprising that under the influence of microwaves, significant changes have occurred in many living systems at different levels from cells to organs. Whereas the body's defense mechanisms can repair the temporal changes in processes in cells and the nervous system, the health effect does not necessarily become evident. Therefore, it is reasonable to discuss biological and health effects separately.

The probability that the defense mechanism can repair the changes caused by long-term exposure to radiofrequency radiation in a living environment is expected to be much smaller compared to that caused by short-term exposure in experimental studies. Further consequences in health, including neuronal disorders, neurodegeneration, and genome instability, the immune system, the male and female reproduction systems, and others are possible.

The results on the biological effects of microwave fields on cells and, consequently, human health are controversial, ranging from negative to zero to positive and beneficial. Therefore, the main applications of microwaves in medicine are discussed.

The living systems are complex and nonlinear and depend on a huge number of variables. The bioelectrical processes in neurons are described by the nonlinear equations (Section 4.5). Therefore, the biological effects are chaotic, and the outcome is unpredictable. In the case of nonlinear chaotic processes, a very small initial change can lead (or not lead) to big alterations, and the prediction of the result becomes impossible (Section 4.5). Therefore, it is complicated to determine the initial reason for the effect of microwave radiation.

The transformation of the biological effects of microwave radiation into health effects is a nonlinear chaotic process (disrupted-line arrows in Figure 5.7). The nonlinearity of the transfer of biological effects into health effects, in turn, makes the impact of microwave radiation on human health even more unpredictable than on biological effects.

This chapter aims to discuss the biological effects of microwave radiation and the possible impact on health, considering the results of published reviews and the knowledge about the nonlinear character of the biological processes (Section 4.5).

7.2 BIOLOGICAL EFFECTS

There is no generally accepted biological theory of the effects of electromagnetic radiation. The intramolecular electric field strengths are higher, 10^6 V/m, and far stronger than the fields from the technical sources of radiofrequency radiation. The intermolecular processes cannot be affected by radiofrequency radiation (Foster, 2000). However, a large variability of experimental results in different biological effects of radiofrequency radiation has been reported over the decades.

The effect of radiofrequency radiation on hydrogen bonds is expected to affect the level of reactive oxygen species (ROS) (Section 4.3). The experimental results have demonstrated an increase in ROS induced by the radiofrequency radiation of different frequencies, exposure durations, and modulation types, as well as in the related cellular-free radical processes in many different biological systems, cell lines, and animal species (Adey et al, 1982; Liu-Liu and Adey, 1982; Sanders et al, 1985; Lai and Levitt, 2022; Hu et al., 2021; Schuermann and Mevissen, 2021). Therefore, the biological effects of radiofrequency radiation cover a wide spectrum of biological effects.

The list of abstracts collected by the leading scientist in biological effects of electromagnetic fields, Professor Emeritus at the University of Washington, Henry Lai, consists of 332 studies published from 1997 to August 2023. The analysis of the database shows that the majority of studies are animal in vivo studies. From the 332 studies, 297 (89%) reported significant effects; 36 (11%) studies found no significant effects. The alterations in cellular-free radical status reported in different studies are the effect of radiofrequency radiation accepted by the majority of researchers. The effect occurs at low levels of radiation: from 82 studies performed at a specific absorption rate of less than 0.4 W/kg, in 79 studies, the effect has been reported. The alterations in ROS and free radicals induced by radiofrequency radiation can occur due to the disturbance of hydrogen bonds by radiofrequency radiation (Section 4.3). Citing Prof. Lai: "These data support the assertion that 'Radiofrequency radiation affects cellular free radical processes.'"

The review of 112 cellular, human, animal, and plant studies with low-level radiofrequency radiation aims at dependencies of the effects on intensity, exposure duration, and modulation (Lai and Levitt, 2022). The authors concluded that the biological outcome of changing the intensity and duration of radiofrequency exposure is basically unpredictable, mainly due to the complex nature of the biological system studied. Similar conclusions have been made based on the analyses of EEG studies (Section 5.1). Only some conclusions about the importance of modulation have been pointed out. The differences occur between continuous-wave and modulated radiation effects depending on the frequencies and types of modulation. The important contribution of modulation is discussed and explained in Section 5.10. Various unpredictable and contradictory outcomes can occur when comparing the effects of continuous-wave and modulated radiation.

The recent review supported by the WHO EMF project results in conclusions different from those of Lai and Levitt regarding the impact of radiofrequency radiation on biomarkers of oxidative stress (Meyer et al., 2024). Fifty-six (56) studies, 45 in vivo and 11 in vitro, were included in the systematic review after eliminating 12,353 publications because they did not meet the criteria defined in the published protocol. The analyzed studies predominantly applied continuous-wave exposures at frequencies in the range of 800–2450 MHz. Only five studies on brain effects have been selected for analysis. The effect on biomarkers for oxidized DNA bases in the rodent brain (five studies, $n = 98$) had an inconsistent effect, varying from a large decrease of −3.40 to a large increase of 2.2. The authors underlined that statistical heterogeneity in the main analysis cannot be attributed to combining different species and the reason for heterogeneity remains unexplained. The authors concluded: "Overall, the effects were inconsistent across studies and there may or may not be an effect of RF-EMF exposure and the certainty of the evidence is low for both directions."

The difference between the conclusions in the two reviews is mainly caused by a different selection of studies for analysis. Lai and Levitt (2022) included studies with modulated and continuous-wave exposures, underlined the importance of modulation, and concluded that unpredictable and contradictory outcomes can occur when comparing the effects of continuous-wave and modulated radiation. Meyer et al. dominantly selected studies using continuous-wave exposure for analysis. Considering the importance of modulation (Section 5.10), it would be said without any experiments and analysis that the effect is of low certainty.

The authors of both reviews emphasize the statistical heterogeneity and unpredictability of the outcome of radiofrequency radiation exposure. The dispersity between the conclusions about the possibility of effects supports the complexity of the effects and demonstrates the chaotic nature of biological systems. The diverse results due to the nonlinear chaotic response of biological systems to the exposure to microwave radiation make the effect unpredictable and difficult to reveal but do not negate the effect.

7.3 EFFECTS OF MILLIMETER WAVES

At frequencies higher than 10 GHz, the dielectric polarization, which is the main source of the action mechanism of weak radio radiation, becomes small, and thus the effect of rotation of dipolar molecules becomes insignificant. Therefore, the effect of microwave radiation related to the rotation of dipolar molecules is significantly reduced. Nevertheless, millimeter waves have shown significant biological effects and efficacy in medical applications at frequencies of 37–78 GHz (Radzievsky et al., 2000, 2001, 2008; Zadobov et al., 2011; Ziskin 2006).

The mechanisms of the biological effects of millimeter waves and the possible consequences for health and their nature are more complex than the effects at lower radiofrequencies. Structural changes in cell membranes have been proposed as an alternative possible mechanism of action (Szabo et al., 2006). The study by Soubere Mahamoudi et al. (2016) directly demonstrates the non-thermal effect of radio radiation at 60.4 GHz (5G NR FR2 frequency) on cell metabolism.

The radiation at higher radiofrequencies does not penetrate deep into the body (Section 3.3). Penetration depth decreases inversely with the square root of frequency. On the one hand, the decrease in penetration depth with frequency protects internal tissues from radiation. The use of higher frequencies reduces the depth of penetration, which is expected to be associated with fewer magnitudes of health effects.

On the other hand, the decrease in penetration depth with frequency leads to an increase in the power density of the absorbed radiation in the skin and an increase in the skin temperature, which becomes most critical for millimeter waves. Calculations at frequencies between 10 GHz and 1 THz showed maximum temperature rise values within 0.2 C for a radiation power density of 10 W/m^2 (Sasaki et al., 2017). Therefore, an increase in skin temperature at radiation power densities below 10 W/m^2 should not increase the likelihood of thermal effects.

Although millimeter waves directly affect the skin, the potential health effects are not limited to the skin. The peripheral nervous system, stimulated by nerve endings in the skin, initiates a chain of physiological processes. Two models have been proposed to explain the possible mechanisms of the effects of millimeter waves: (1) direct activation of skin cells induces the secretion of molecular signaling factors into the general circulation, and (2) stimulation of the peripheral nervous system activates the central nervous system (CNS) and induces the secretion of opioid peptides

(Radzievsky et al., 2001; Zhabodov et al., 2011). The important role of the peripheral nervous system in the effect of millimeter waves is confirmed by experiments that showed that the radiation causes a stronger effect when skin areas with a high concentration of nerve endings are exposed (Radzievsky et al., 2000).

The mechanisms of action of millimeter waves and related possible biological effects are still not clear. Recently published reviews highlight that there is insufficient research on the higher frequencies of 24–100 GHz (Simko and Mattson, 2019; Thielens, 2021).

7.4 EFFECT ON CANCER

The main attention in the evaluation of the radiofrequency radiation health effects has been directed to the risk of tumors. Radiofrequency electromagnetic radiation has been classified as possibly carcinogenic to humans (Group 2B) by the International Agency for Research on Cancer (IARC) (IARC, 2013). According to IARC, animal studies, as well as epidemiological ones, showed limited evidence of carcinogenicity.

The disturbance of hydrogen bonds (Section 4.3), keeping together DNA molecules by radiofrequency radiation, indicates a possibility that radiofrequency radiation can be carcinogenic. However, due to the long latent period, diverse exposure conditions, and biological parameters, the causal relationship is difficult to reveal.

7.4.1 Epidemiologic studies

There are still no consistent experimental or epidemiologic results of carcinogenicity of radiofrequency radiation in humans. Experimental investigations of cancer in humans are hardly possible. In population-based epidemiologic studies, it is not possible to find a control group in the modern world covered fully by radiofrequency radiation. Many confounding factors can be carcinogenic, like chemicals in the air, food, household products, tobacco, and alcohol. Considering the impact of confounders is difficult and partly even impossible, as sometimes people may not be aware of them.

The investigation of the dependence of cancer cases on the level and duration of exposure would provide some knowledge. However, due to the variations of individual sensitivity to the radiation (Section 5.6) and the uncertain threshold of the effect (Section 4.11), the results show a rather distorted picture. The analysis of the number of cancer diagnoses over decades reflects rather the development of medical equipment and methods for the detection of cancer.

Due to the reasons above, population-based studies in exposed humans have provided rather inconsistent results and different conclusions. No conclusive evidence of an association between the use of mobile and cordless phones and meningioma was found (Carlberg et al., 2013). The authors concluded that an indication of increased risk was seen in the group with the highest cumulative use but was not supported by statistically significant increasing risk with latency. On the contrary, the recent meta-analyses of case-control studies found that cellular phone use with a cumulative call-time of more than 1000 h statistically significantly increased the risk of tumors (Choi et al., 2020). However, the later review concludes that claims of a radiofrequency radiation cancer link are not supported by the current evidence base, whereas much of the research to date, however, has been undermined by methodological shortcomings (Grimes 2022). A group of scientists from different countries, based on the accumulated evidence, recommend that IARC re-evaluate its 2011 classification of the human carcinogenicity of radiofrequency radiation (Miller et al., 2019).

7.4.2 Animal Studies

More consistent are the results of animal studies that support IARC's conclusions of a carcinogenic potential of radiofrequency radiation. The results of the long-term National Toxicology Program (NTP, 2018) on toxicology and carcinogenesis studies in Sprague–Dawley rats exposed to whole-body radiofrequency radiation at the frequency of 900 MHz and modulations GSM and CDMA used by cell phones were published in 2016. During two-year toxicology and carcinogenicity studies of male and female rats and mice, 90 animals per group in 20 special reverberation chambers were exposed to whole-body radiofrequency radiation for ~9 hours per day (18 hours 20 minutes per day in 10 minutes on/10 minutes off cycles) at specific absorption rate (SAR) levels of 1.5, 3, and 6 W/kg. The results showed that, in general, survival was greater in all groups of GSM or CDMA radiofrequency radiation-exposed rats compared to controls. There was a significant SAR-dependent trend for increased gliomas in the brains of rats exposed to CDMA-modulated radiofrequency radiation. Increased incidence of schwannoma was observed in the hearts of male rats. No exposure-related effects were observed in the brains or hearts of female rats. Tumor types observed in this study are similar to those observed in some epidemiology studies of cell phone users.

The NTP studies performed on two types of animals indicated that the effects of radiofrequency radiation affect each animal species differently. Rats and mice are biologically very similar animals, and differences in their responses to radiofrequency radiation are difficult to understand. However, the differences in the results between rats and mice would be related more to technical issues. Animal exposure levels are reported in the NTP study as whole-body SAR, and the corresponding electric field strengths are not indicated in the report. Mice are much smaller than rats, and the field power density required for achieving equal SAR values in rats and mice (as well as the corresponding electric field strength) should be much lower in mice experiment than in experiments with rats. At a much weaker field strength in mice experiments, the effect of radiofrequency radiation cannot become evident. The NTP report indicates a significant SAR-dependent sixth-degree polynomial positive trend. However, the polynomial SAR-dependent trend of the effect is nonlinear and does not exclude but rather supports the field-dependent trend. The finding that radiofrequency radiation effects depend rather on field strength, not on SAR, supports the nonthermal nature of the effects. The trend that the effect level follows electric field strength, not power density, is also evident in EEG studies (Section 5.7).

The NTP study on rats has been partly replicated in the Ramazzini Institute by Falcioni et al. (2018). A life-span carcinogenic study was performed to evaluate the carcinogenic effects of radiofrequency radiation. In the largest long-term study, 2448 male and female Sprague–Dawley rats were exposed from prenatal life until natural death to the 1.8 GHz GSM far field of 0, 5, 25, 50 V/m with a whole-body exposure for 19 h/day. A statistically significant increase in the incidence of heart Schwannomas was observed in treated male rats at the highest dose (50 V/m). Furthermore, an increase in the incidence of heart Schwann cell hyperplasia was observed in treated male and female rats at the highest dose (50 V/m), although this was not statistically significant. An increase in the incidence of malignant glial tumors was observed in treated female rats at the highest dose (50 V/m), although this was not statistically significant. The findings of far-field exposure to radiofrequency radiation are consistent with and reinforce the results of the NTP study on near-field exposure, as both reported an increase in the incidence of tumors of the brain and heart in exposed Sprague–Dawley rats.

Tumor types observed in NTP and Ramazzini Institute studies are similar to those observed in some epidemiology studies on cell phone users. These experimental studies support the carcinogenic potential of radiofrequency radiation in humans. However, the results of animal studies may not be directly transferable to humans.

7.4.3 Cellular Studies

Substantial systematic reviews on radiofrequency radiation effects in cancer biomarkers of human and animal cells provide a comprehensive picture of the situation (Meyer et al., 2024; Romeo et al., 2024; Simko et al., 2025).

A possible genotoxicity of microwave radiation related to oxidative stress was assessed by Meyer et al. (2024). Fifty-six (56) studies, 45 in vivo and 11 in vitro, in which cells (in vitro) or animals (in vivo) were exposed to frequencies in the range 800–2450 MHz, were included in the systematic review. Of 56 studies, 52 studies with 169 individual results were included in the meta-analysis. In in vitro studies in human cells (three studies, $n = 110$), there were inconsistent increases in biomarkers for oxidized DNA bases, where the effects varied between 0.01 and 7.12 in four results (two of them being statistically significant). In rodent cells (three studies, $n = 24$), there was not a statistically significant large effect (2.07) in biomarkers for oxidized DNA bases. The biomarkers for modified proteins and amino acids in human cells (one study, $n = 18$) showed a large effect of 1.07. In rodent cells (two studies, $n = 24$) a medium effect of 0.56 was observed. The authors concluded that the relation between exposure to radiofrequency radiation and biomarkers of oxidative stress was of very low certainty.

A systematic review of the scientific literature on genotoxic effects induced by radiofrequency radiation at levels lower than ICNIRP guidelines in in vitro experimental models was performed by Romeo et al. (2024). Out of 7750 original studies, 159 articles were eligible for inclusion. From the extracted data, 1111 experiments (defined as independent specific combinations of diverse biological and electromagnetic parameters) were identified and analyzed. The large majority (80%) of experiments reviewed did not show statistically significant genotoxic effects of radiofrequency exposures, and most "positive" studies were rated as of moderate to low quality. Some of the experiments reporting genotoxic effects did not confirm the original results when repeated by another study by the same team.

Some other experimental results reporting effects have not been replicated by independent research groups, and the results of experiments were not confirmed by other research. The authors concluded that in the assessment restricted to studies reporting a significant effect of the exposure on the outcome, an overall assessment reaches "low" confidence in the evidence that radiofrequency radiation induces genotoxic effects in mammalian cells. However, 80% of the experiments reviewed showed no effect of exposure on the large majority of endpoints independently of the exposure features, level, and duration (moderate evidence of no effect). The authors conclude that the analysis of the papers included in this review suggests that radiofrequency exposure does not increase the occurrence of genotoxic effects in vitro.

The review by Simko et al. (2025) is aimed at the assessment of evidence needed to determine whether exposure to radiofrequency electromagnetic fields, below the levels recommended in the ICNIRP (2020) guidelines, can influence any of the ten key characteristics of human carcinogens developed by the

IARC. The ten key characteristics are as follows:

1. Is electrophilic or can be metabolically activated to an electrophile key characteristics and their relevance to carcinogenesis;
2. Is genotoxic;
3. Alters DNA repair or causes genomic instability;
4. Induces epigenetic alteration;
5. Induces oxidative stress;
6. Induces chronic inflammation;
7. Is immunosuppressive;
8. Modulates receptor-mediated effects;
9. Causes immortalization;
10. Alters cell proliferation, cell death, or nutrient supply.

The authors performed the analyses for six criteria and used the results of the assessment of genotoxicity and oxidative stress from Romeo et al. (2024) and Meyer et al. (2024).

On the assessment of the other 8 criteria, 119 in vitro and 40 in vitro measurements of in vivo studies were identified, with 38% reporting statistically significant effects of exposure. We identified a strong association between the quality of study and the outcome, with those meeting more criteria less likely to report statistically significant effects. Effects were reported over the entire frequency range, exposure levels, and biological endpoints, with no apparent pattern of exposure parameters resulting in effects. Only criterion 10 (alters cell proliferation, cell death, or nutrient supply) has sufficient studies to analyze, but the other criteria had few studies and diverse endpoints. A few relatively high-quality positive studies require follow-up through additional targeted studies. The heterogeneity and overall poor study quality suggest the need for high-quality studies on these endpoints.

All authors underline that the relationship between radiofrequency radiation and biomarkers is of very low certainty due to high heterogeneity in studies and their low quality caused by flaws in methods like inaccurate measurements of exposure and biomarkers, incomplete statistical evaluation of results, and missing information on the blinding of research.

7.5 NEUROPSYCHOLOGICAL EFFECTS

Less attention in the evaluation of microwave radiation health effects has been directed to neurological diseases and mental disorders. The analysis of the studies in resting-state EEG, event-related potentials, and behavior over the last 18 years (Section 5.1) shows that low-level radiofrequency radiation causes alterations in human EEG and behavior. Microwave radiation-related neurophysiologic alterations can cause neurological and mental disorders not necessarily related to alterations in oxidative stress or cells.

The relationship between the use of mobile communication and neurophysiologic disorders is supported by epidemiologic studies. The cohort study on cellular telephones and CNS disease risks was conducted on a large nationwide cohort study of 420,095 persons in Denmark whose first cellular telephone subscription was between 1982 and 1995, who were followed through 2003 for hospital contacts for a diagnosis of a CNS disorder (Schüz et al., 2009). Standardized hospitalization ratios (SHRs) were increased by 10–20% for migraine and vertigo. No associations were seen for amyotrophic lateral sclerosis, multiple sclerosis, or epilepsy in women. SHRs decreased by 30–40% and were observed for dementia (Alzheimer's disease and vascular and other dementia), Parkinson's disease, and epilepsy among men (Schuz et al., 2009).

A prospective cohort study to assess the association between the amount of mobile phone use at baseline and frequency of headache, tinnitus, or hearing loss at four-year follow-up was performed by Auvinen et al. (2019). The mobile phone call data from operator records of mobile phone subscriptions and filled questionnaires in both baseline and follow-up on symptoms and potential confounders were analyzed for n = 21,049 participants in Sweden and n = 3120 in Finland. The participants with the highest decile of recorded call-time (average call-time >276 minute per week) at baseline showed a weak, suggestive increased frequency of weekly headaches at four-year follow-up (adjusted odds ratio 1.13, 95% confidence interval 0.95–1.34). There was no obvious gradient of weekly headache with increasing call-time (p trend 0.06). The association of headache with call-time was stronger for the UMTS network than older GSM technology, despite the latter involving higher exposure to radiation. Tinnitus and hearing loss showed no association with call-time (Auvinen et al., 2019).

The alterations in EEG caused by radiofrequency radiation are similar in the majority of the studies: an increase in alpha, beta, and gamma band levels in EEG, as well as higher complexity of the signal (Hinrikus et al., 2021). Similar changes in EEG are characteristic of major depressive disorder (Ahmadlou et al., 2012; Bachmann et al., 2018a; Fingelkurs, Fingelkurs, 2015; Hosseinifard et al., 2013; Knott et al., 2001). These observations are based on the results obtained by many authors and can hardly be a coincidence.

The causality between radiofrequency radiation and depression is not clear. The results of the study based on a special questionnaire indicated an association between higher mobile phone use and symptoms of depression (Thomee et al., 2011). The recent Australian study performed on 655 adults has found an association between smartphone use and depression, stress, anxiety, and sleep quality (Khan et al., 2023). However, it is complicated to differentiate between the direct effect of radiofrequency radiation and psychological factors related to the high use of mobile phones, leading sometimes even to addiction (De-Sola Gutiérrez et al., 2016).

Depression has become a common mental disorder over the last few decades, with the highest prevalence among individuals aged 18–25 (13.1%) (NIH, 2021). The use of mobile telecommunication technology has become more intense, with prevalence among young people during the same period. Is this a coincidence or causality? The use and addiction to smartphones have a psychological impact on the user as well as a physical impact related to the radiation. It is highly complicated to differentiate their contribution to mental health.

The experimental results demonstrate that at the same parameters of radiofrequency radiation exposure and identical conditions, only a part of people are affected (Section 5.6). The rate of sensitive people varies from 13% to 31% depending on the modulation frequency. It is not clear whether some people are "immune" to the radiation permanently or occasionally. It is common knowledge that the sensitivity to various external biological and chemical factors, like viruses or chemicals, differs between individuals. Some people may be more sensitive and even allergic to certain factors. This also explains the so-called hypersensitivity to radiofrequency radiation.

The results of some studies indicate that the changes related to microwave radiation in the neuronal system (EEG signal) are much more frequent than in subjective behavior (Vecsei et al., 2013). The exposure-induced effects have been seen in objective indicators (EEG, glucose metabolism) but not in cognitive performance (Section 6.2). In the EEG experiments performed with 450 MHz 1.6 V/m modulated radiation on about 70 persons in total, none of the subjects have reported any feelings under the radiation and cannot differentiate between exposed and sham sessions (Hinrikus et al., 2008; Rodina et al., 2005). The situation where objective physiological symptoms appear earlier than subjective feelings and symptoms is quite usual in the case of many diseases. People would not mention existing mild signs of the microwave radiation effect.

7.6 MICROWAVE HEARING

A high-level pulsed radiofrequency radiation can create a loud buzzing burst of sound in the human head. People can directly "hear" pulsed microwave radiation. The microwave auditory effect was first described by Frey (1962) and explained as direct electric excitation of the cochlea. The detailed thermoelastic theory of microwave energy to sound transduction has been developed by Professor James C. Lin from the University of Illinois (1978, 2021a, 2022). Absorption of microwave pulses causes rapid extension in the head tissues and creates an acoustic pressure wave in the inner ear that can be detected by the cochlea as a sound. Microwave hearing doesn't go through the ear; it goes directly from the head tissue to the inner ear cochlea.

The auditory effect of microwave radiation is thermal. The effect occurs from a rapid (μs) minimal rise of temperature (10–6°C) in the head produced by the absorption of pulsed microwave radiation. The threshold of the effect is determined by the threshold of auditory sensations to pressure waves created by absorbed microwave energy. The threshold depends on microwave frequency and pulse duration and varies between about 6 and 50 kW/m^2 (Lin, 2022a). Due to the short duration of high-power microwave pulses, the radiation does not generate noticeable total amounts of heat in the brain. The microwave auditory effects have been reported at frequencies from 200 MHz to 3 GHz. The microwave auditory effect has been discussed as a possible explanation for Havana syndrome (Lin 2021b).

7.7 HAVANA SYNDROME

Havana syndrome, named after the place where it was first reported, is characterized by various symptoms like vertigo, headaches, tinnitus, pain, strange sounds, and visual and cognitive problems (Golomb, 2018). The Havana syndrome, also known as anomalous health incidents (AHIs), is considered a disease in the United States that needs treatment.

First, the American diplomatic personnel in Havana, Cuba, reported symptoms in 2016. Beginning in 2018, diplomatic personnel around the world, including Moscow, Tbilisi, Poland, Guangzhou, Taiwan, Australia, Hanoi, China, Washington, and other places, have reported similar symptoms. People have reported symptoms in a variety of locations, including embassies, hotels, and homes.

Several research teams and commissions initiated by various institutions, including the US government, FBI, the State Department, US intelligence, the National Academy of Sciences, and others, have been involved in the investigation of the Havana syndrome (Asadi-Pooya, 2023; National Academies of Sciences, Engineering, and Medicine, 2020).

At the request of the US government, the researchers from the University of Pennsylvania examined the diplomats from the US government personnel in Havana (Swansson et al., 2018). The preliminary report describes the syndrome as a concussion of a retrospective case series, persistent cognitive, vestibular, and oculomotor dysfunction, sleep impairment, and headaches, associated with directional audible and/or sensory phenomena of unclear origin. The conclusion was that individuals have sustained injury to widespread brain networks without an associated history of head trauma (Swanson et al., 2018). The later study by Verma et al. (2019) included 40 US government personnel and 48 healthy controls. The used brain MRI revealed significant between-group differences in whole brain white matter volume, regional gray and white matter volume, cerebellar tissue microstructural integrity, and functional connectivity in the auditory and visuospatial subnetworks but not in the executive control subnetwork. The clinical importance of these differences is uncertain and may require further study.

The recent results of NIH AHI Intramural Research Program exploratory studies included 86 participants reporting AHI and 30 matched control participants. The results of the MRI study revealed no significant differences in imaging measures of brain structure or function between individuals reporting AHIs after correction for multiple comparisons. These findings suggest that the origin of the symptoms of participants with AHIs may not be linked to an MRI-identifiable injury to the brain

(Pierpaoli et al., 2024). No clear correlation between imaging and clinical variables was evident. There were also no significant differences in most tests of auditory, vestibular, cognitive, visual function, or blood biomarkers between the groups (Chan et al., 2024). However, AHI participants declared significantly worse self-reported and objective measures of balance and had significantly increased symptoms of fatigue, posttraumatic stress disorder, and depression compared with the control participants. Of AHI patients, 24 participants (28%) presented with functional neurological disorders (Chan et al., 2024). The results of MRI studies indicate uncertainty about the medical mechanism of the Havana syndrome.

The causes and the mechanism of the Havana syndrome are still unknown. The idea that microwave radiation is a main cause underlying the Havana syndrome arose in 2018. The high-power pulsed RF/microwave energy can create a rapid thermal expansion and lead to stress waves within the body, which can result in the microwave auditory effect (Section 7.6). In 2018, Columb suggested a pulsed microwave radiation auditory Frey effect as an explanation for the symptoms of Havana syndrome. The author concluded, "Reported facts appear consistent with pulsed RF/MW as the source of injury in affected diplomats." However, the threshold of the auditory microwave effect is very high. The discussion of the microwave auditory effect as a possible reason for Havana syndrome has concluded that "these peak power densities are close to and encompass the 23.8 GW/m^2 value for the dielectric breakdown of air" (Lin, 2021b). The modeling data support the need for high-power pulses required to generate thermal stress waves in the brain (Dagro et al., 2021). The presence of such pulses in a living environment is unlikely. Unfortunately, the author found no information in multiple records about the level of radiofrequency radiation in the buildings where the syndrome was evident.

The thermal auditory microwave effect is the only microwave biological effect so far generally accepted by the scientific community. Therefore, as expected, only the possibility of the thermal effect has been evaluated (Dagro et al., 2021; Foster et al., 2025; Lin, 2021b). In principle, the chance of low-level microwave effects exists. There is a possibility of affecting human emotions and feelings, exciting relevant neuronal frequencies in the brain. Parametric resonance makes it possible to excite the selected frequencies employing minimal field power density (Section 4.9). The microwave brain stimulation is possible (Section 7.8.4). However, our experimental results on the effects of low-level microwave radiation performed in total in more than a hundred subjects in different studies have demonstrated that the short-term radiation causes no feelings or inconveniences during exposure and later (Chapter 6). Our experiments were performed with microwave frequency modulated at different modulation frequencies, including 7 and 40 Hz, expected to be most effective in neurophysiology. The results of our experiments do not support the hypothesis that long-term exposure causes some subjective effects. The experimental data show that the brain physiology instead adapts to the weak radiation, and the alterations in EEG decrease over time. Although the modulated microwave radiation can stimulate the brain (Section 7.8.4), the codes of the neuronal oscillations related to predetermined emotions or feelings are unknown. There is no reason to think that somebody has the knowledge or equipment to affect human emotions and behavior using radiofrequency radiation today. Microwave radiation is unlikely to be the cause of the Havana syndrome.

7.8 APPLICATIONS IN MEDICINE

7.8.1 Electrotherapy (diathermy)

High-frequency electromagnetic fields have been used for therapeutic purposes for more than a century. Interestingly, the first application of high-frequency electromagnetic waves was related to medicine; the application in telecommunication was demonstrated somewhat later (Section 2.1). Jaques-Arsene d'Arsonval (1851–1940) studied a wide variety of the physiological effects of alternating currents, time-varying electric and magnetic fields, and high frequency. From 1891 to 1893 he studied the effects of muscle stimulation, pulse changes, perspiration, and nervous stimulation. He found a positive effect of a high-frequency electric field of about 5 MHz frequency and introduced the use of high-frequency currents. The current is now known as the d'Arsonval current. The medical terms "darsonvalization" and "diathermy" used in physiotherapy originated from his research in 1892. He is the founder of the field of electrophysiology and electrotherapy, known as diathermy in physiotherapy.

Traditional heating has mainly a surface effect, and heating of inner tissues occurs only due to thermal conductivity between tissues and is not effective. A radiofrequency field penetrating tissues provides deep heating of tissues and inner organs. Diathermy treatment increases blood flow and improves microcirculation. It helps relax muscles and joints, reduces regenerative processes and swelling, and has analgesic and anti-inflammatory effects.

Diathermy is based on the thermal radiofrequency effect. The rise in temperature increases diffusion and so accelerates all processes related to the movement of molecules and ions. Considering the coherent nature of technical electromagnetic radiation and the experience of microwave chemistry, the effect is expected to be stronger compared to traditional thermal heating (Section 5.6).

Millimeter-wave therapy uses low-intensity electromagnetic radiation at the most common frequencies, 42.2, 53.6, and 61.2 GHz (Ziskin, 2006). The localized area of the skin is exposed to low-level radiation without perceptible heating and rare minor side effects. Millimeter-wave therapy has demonstrated good results in the treatment of a large variety of diseases, including cardiovascular disorders, diabetes, dermatitis, gastrointestinal disorders, wound healing, pain relief, and the reduction of toxic side effects of chemotherapy in cancer patients.

7.8.2 Hyperthermia in cancer treatment

Hyperthermia has been proven to be an effective method in the treatment of cancer (Field,1976; Hahn, 2012; Mallory et al., 2016). It employs the fact that the properties of cancer cells differ from those of normal cells. Cancerous cells are already damaged cells and therefore much less able to tolerate the stress of heat than normal cells. Additionally, the electrical properties of cancer cells differ from those of normal cells, and thus they absorb more electromagnetic energy than normal cells. This provides an opportunity to use targeted radiation to heat and destroy cancer cells with minimal damage to healthy tissue.

Hyperthermia uses higher temperatures compared to electrotherapy. Three temperature intervals are in use. Mild hyperthermia employs a temperature close to that of a natural fever level of 39–40°C. It is a whole-body treatment and stimulates natural immunologic processes.

Moderate hyperthermia uses higher temperatures of 40–42°C. It is used as localized heating of a selected organ or body area. Moderate hyperthermia can damage cells directly. Additionally, the heating increases the sensitivity of cells to radiotherapy and chemotherapy.

High-temperature hyperthermia uses temperatures of 43–45°C. High-temperature heating is targeted hyperthermia to destroy cancerous cells in a determined location. Special combinations of antennas provide focusing of microwave radiation on a relevant point. If the tumor is located in the inner organs, special needle-like antennas are used. In some cases, microwave ablation at a temperature of 50°C and higher is used to destroy cancerous cells. The precise location of cancerous cells can be detected using microwave radiometry.

7.8.3 Microwave radiometry in medical imaging

Microwave radiometry is a passive method of medical imaging applicable to the detection of cancerous cells (Carr, 1989; Fisher et al., 2022; Hinrikus et al., 1998; Hinrikus and Riipulk, 2006).

A radiometer is a sensitive device that measures the radiation temperature of objects. Each body with a temperature higher than 0 K radiates thermal electromagnetic radiation. The spectral power density of the radiation $P(f)$

$$P(f) = \alpha \cdot kT \tag{7.1}$$

where α is the radiation coefficient equal to the absorption coefficient, k is the Boltzmann constant, and T is the absolute temperature. The radiometric method is passive and non-invasive, not using chemicals to detect damaged cells. The spatial resolution of the method is determined by the wavelength and is about a millimeter. The radiometric method is more effective in a homogeneous medium and is successfully applied in the early detection of breast cancer.

The electric properties determining the radiation coefficient α and the temperature of cancerous cells differ from those in normal tissues. The radiometric method can detect even a few cancerous cells in an early stage of cancer. In 1989, Kenneth Carr received the Application Award of the IEEE Microwave Theory and Techniques Society for the development of the microwave radiometric system for the detection of cancer (Carr, 1989). Various methods applying multiple detectors have been developed to improve spatial resolution (Hinrikus and Riipulk, 2006). The method has been demonstrated to be useful in homogeneous tissues and breast cancer detection.

7.8.4 Brain stimulation

Electric stimulation is based on the direct effect of the electromagnetic field and differs from electrotherapy using thermal effects and heating by the electromagnetic field. Electric stimulation has become a widely used treatment method in neurology. To provide deep penetration, pulse-modulated direct magnetic or electric fields are used. The parameters of pulses are selected inside the spectrum of the brain's natural oscillations. The most widely used non-invasive methods are repetitive transcranial magnetic stimulation (rTMS) (Lefaucheur et al., 2020; Rachid 2018) and transcranial direct current stimulation (tDCS) (Groppa et al., 2010; Mordillo et al., 2017; Thair et al., 2017). The most common invasive method is deep brain stimulation (Blomstedt and Hariz, 2010; Camacho-Conde et al., 2022). Brain stimulation by radiofrequency electromagnetic radiation is a promising method that is not yet in clinical use (Bachmann et al., 2018b; Perez et al., 2021).

Repetitive transcranial magnetic stimulation uses a magnetic field produced by a coil that provides targeted excitation of the brain areas of interest. The mechanism of rTMS is based, according to Ampère's law (Eq. 1.6), on the impact of the magnetic field on moving ions. The changes in ions' movement result in alterations of the membrane potential and excitability of the cerebral cortex. The stimulation can affect cerebral blood flow and oxygen consumption.

Transcranial direct current stimulation uses low-level electric current through electrodes and produces non-targeted excitation of the brain. It is possible to direct the current to different parts of the brain with the appropriate placement of the stimulation electrodes. The movement of ions causes alterations in membrane potential and synaptic transmission. tDCS produces changes in cortical activity and functional connectivity between different brain regions.

Deep brain stimulation uses implanted electrodes placed in damaged areas of the brain. The stimulating electric pulses are produced and controlled by a pacemaker-like device placed under the skin in the upper chest. The electric field affects ions' movement and can cause excitation in neighboring axons, increase local cerebral blood flow, and stimulate astrocytes to release calcium.

The mechanism of brain stimulation using low-level modulated radiofrequency radiation is based on the effect of parametric excitation of the brain neuronal oscillations (Section 5.9). Mental disorders cause alterations in the EEG spectrum, decreasing or increasing some specific frequencies. The correction of the spectrum exciting appropriate EEG frequencies is expected to decrease the symptoms of mental disorders. The results of the performed study using the radiofrequency radiation of 450 MHz modulated at 40 Hz and a power density of 0.16 mW/cm^2 suggest that low-level modulated radiation can cause excitation of EEG rhythms at the predicted frequencies (Bachmann et al., 2018b).

An increase in the level of extracellular Amyloidβ (Aβ) is a primary cause of dementia and Alzheimer's disease. The recent study by Perez et al. (2021) has demonstrated that exposure to the low-level (SAR < 0.4 W/kg), pulse-modulated radiofrequency radiation of 64 MHz frequencies can significantly reduce the Aβ levels in primary human brain cultures without secondary toxicity. This finding is of interest and potentially promising in clinical applications.

The neurophysiological mechanisms of brain stimulation methods are still not clear. The changes in brain activity have been demonstrated from single neurons to local field potential and to brain-wide cortical network effects. The theory combines a synthesis of synaptic and network effects across scales of observation, including micro-, meso-, and macroscale circuit alterations (Neuman et al., 2023).

Brain stimulation has become one of the most acceptable therapeutic approaches in recent years and a promising tool in the treatment of Parkinson's disease, Alzheimer's disease, vascular dementia, major depressive disorder, attention-deficit hyperactivity disorder, autism spectrum disorder, and others (Camacho-Conde et al., 2022).

7.9 CONCLUSIONS

Very different biological effects of low-level microwave radiation have been reported by hundreds of scientists. A remarkable part of the performed studies has indicated significant effects. Despite that, no regular relationship between the intensity, frequency, and other parameters of the radiation was revealed; only the importance of modulation was underlined.

The transfer of biological effects to impact health is a complex, nonlinear process. Therefore, the effects of low-level microwave radiation on health are unpredictable and of low certainty. Anyway, the possibility of microwave radiation health effects cannot be neglected.

- The complex nonlinear nature of the microwave radiation effects is the reason why the interpretation of the results of experimental studies, despite decades of investigations, is controversial and debatable. Further investigation, in principle, cannot contribute significantly to solving the problem.
- The missing regular dependence on the level and character of exposure reported by the majority of authors supports the nonlinear nature of the effect.
- The analyses of experimental studies show a big difference between the rate of effects related to modulated and continuous wave exposures. Modulation is important to reveal the effects.
- People cannot feel or mention the mild impact of microwave radiation on their nervous system.
- The medical applications of microwave radiation in clinical use mainly employ thermal effects.
- The promising opportunities of low-level microwave radiation in the treatment of neurodegeneration need further investigation.
- The health effects are of low certainty and are often claimed not to have been proven. But the lack of health effects has not been proven either. Given the biological effects, it is essentially impossible to deny the impact on health.

REFERENCES

Adey WR, Bawin SM, Lawrence AF, (1982). Effects of weak amplitude-modulated microwave fields on calcium efflux from awake cat cerebral cortex. *Bioelectromagnetics*. 3(3):295–307

Ahmadlou M, Adeli H, Adeli A, (2012). Fractality analysis of frontal brain in major depressive disorder. *Int J Psychophysiol.* 85:206–211.

Asadi-Pooya AA, (2023). Havana syndrome: a scoping review of the existing literature. *Rev Environ Health.* Aug 15;38(4):655–661. doi: 10.1515/reveh-2021–0182.

Auvinen A, Feychting M, Ahlbom A, Hillert L, Elliott P, Schüz J, Kromhout H, Toledano MB, Johansen C, Poulsen AH, Vermeulen R, Heinävaara S, Kojo K, Tettamanti G, COSMOS Study Group, (2019). Headache, tinnitus and hearing loss in the international Cohort Study of Mobile Phone Use and Health (COSMOS) in Sweden and Finland. *Int J Epidemiol.* 48(5):1567–1579.

Bachmann M, Päeske L, Kalev K, Aarma K, Lehtmets A, Ööpik P, Lass J, Hinrikus H, (2018a). Methods for classifying depression in single channel EEG using linear and nonlinear signal analysis. *Comput Methods Programs Biomed.* 155:11–17.

Bachmann M, Lass J, Ioannides, AA, Hinrikus H, (2018b). Brain stimulation by modulated microwave radiation: a feasibility study. *EMF-Med 1st World Conference on Biomedical Applications of Electromagnetic Fields (EMF-Med).* doi: 10.23919/EMF-ED.2018.8526055.

Barnes F, Greenebaum B, editors. (2018). *Biological and medical aspects of electromagnetic fields.* CRC Press, Taylor & Frances, Boca Raton, FL.

Blomstedt P, Hariz MI, (2010). Deep brain stimulation for movement disorders before DBS for movement disorders. *Parkinsonism Relat Disord.* 16(7):429–433.

Camacho-Conde JA, Gonzalez-Bermudez MDR, Carretero-Rey M, Khan Z, (2022). Brain stimulation: a therapeutic approach for the treatment of neurological disorders. *CNS Neurosci Ther.* 28(1):5–18. doi: 10.1111/cns.13769.

Carlberg M, Söderqvist F, Mild HK, Hardell L, (2013). Meningioma patients diagnosed 2007--2009 and the association with use of mobile and cordless phones: a case-control study. *Environ Health.* 12(1):60. doi: 10.1186/1476-069X-12–60.

Carr, K, (1989). Microwave radiometry: its importance to the detection of cancer. *IEEE Trans Microwave Theory Tech.* MTT37:1862–1869.

Chan L, Hallett M, Zalewski CK, Brewer CC, Zampieri C, Hoa M, Lippa, SM, Fitzgibbon, E, French LM, Moses AD, van der Merwe AJ, Pierpaoli C, Turtzo LC, Yonter S, Shahim P, NIH AHI Intramural Research Program Team, (2024). Clinical, biomarker, and research tests among United States government personnel and their family members involved in anomalous health incidents. JAMA. Published online March 10, 2024. doi: 10.1001/jama.2024.2413.

Choi YJ, Moskowitz JM, Myung SK, Lee YR, Hong YC, (2020). Cellular phone use and risk of tumors: systematic review and meta-analysis. *Int J Environ Res Public Health.* 17(21):8079.

Dagro AM, Wilkerson JW, Thomas TP, Kalinosky BT, Payne JA, (2021). Computational modeling investigation of pulsed high peak power microwaves and the potential for traumatic brain injury. *Sci Adv.* 7(44):eabd8405. doi: 10.1126/sciadv.abd8405.

De-Sola Gutiérrez J, Rodríguez de Fonseca F, Rubio G, (2016). Cell-phone addiction: a review. *Front Psychiatry.* 24(7):175.

Falcioni L, Bua L, Tibaldi E, Lauriola M, De Angelis L, Gnudi F, Mandrioli D, Manservigi M, Manservisi F, Manzoli I, Menghetti I, Montella R, Panzacchi S, Sgargi D, Strollo V, Vornoli A, Belpoggi F, (2018). Report of final results regarding brain and heart tumors in Sprague-Dawley rats exposed from prenatal life until natural death to mobile phone radiofrequency field representative of a 1.8 GHz GSM base station environmental emission. *Environ Res.*165:496–503. doi: 10.1016/j.envres.2018.01.037.

Field SB; (1976). Hyperthermia in the treatment of cancer. *Cancer Treat Rev.* 6:63–94.

Fingelkurts AA, Fingelkurts AA, (2015). Altered structure of dynamic electroencephalogram oscillatory pattern in major depression. *Biol. Psychiatry.* 77:1050–1060.

Fisher L, Fisher O, Chebanov D, Vesnin S, Goltsov A, Turnbull A, Dixon M, Kudaibergenova I, Osmonov B, Karbainov S, Popov L, Losev A, Goryanin I, (2022). Passive microwave radiometry and microRNA detection for breast cancer diagnostics. *Diagnostics (Basel).* 13(1):118. doi: 10.3390/diagnostics13010118.

Foster KR, (2000). Thermal and non-thermal mechanisms of interaction of radio-frequency energy with biological system. *IEEE Trans Plasma Sci.* 28:15–23.

Foster KR, Diao Y, Chou C-K, (2025). Acoustic pressures in the head from pulsed microwaves: can they explain the Havana Syndrome? https://www.techrxiv.org/users/952786/articles/1322094

Frey AH, (1962). Human auditory system response to modulated electromagnetic energy. *J Appl Physiol.* 17:689–692.

Golomb BA, (2018). Diplomats' mystery illness and pulsed radiofrequency/microwave radiation. *Neural Comput.* 30(11):2882–2985.

Grimes DR, (2022). Radiofrequency radiation and cancer: a review. *JAMA Oncol.* 8(3):456–461.

Groppa S, Bergmann TO, Siems C, Mölle M, Marshall L, Siebner HR, (2010). Slow-oscillatory transcranial direct current stimulation can induce bidirectional shifts in motor cortical excitability in awake humans. *Neuroscience.* 166(4):1219–1225.

Hahn GM, (2012). *Hyperthermia and cancer*. Springer, New York.

Hinrikus H, Riipulk J, Neemela T, Podra H, (1998). Sensitivity of microwave radiometer for tissue screening. *Proc Estonian Acad Sci Eng*. 4: 165–177.

Hinrikus H, Riipulk, J, (2006). Microwave imaging. In: M Akay, editor. *Wiley encyclopedia of biomedical engineering*, Wiley, Hoboken, New Jersey, pp. 2329–2340.

Hinrikus H, Bachmann M, Lass J, Karai D, Tuulik V, (2008). Effect of low frequency modulated microwave exposure on human EEG: individual sensitivity. *Bioelectromagnetics*. 29(7):527–538.

Hinrikus H, Lass J, Bachmann M, (2021). Threshold of radiofrequency electromagnetic field effect on human brain. *Int J Radiat Biol.* 97(11):1505–1515.

Hosseinifard B, Moradi MH, Rostami R, (2013). Classifying depression patients and normal subjects using machine learning techniques and nonlinear features from EEG signal. *Comput Methods Programs Biomed.* 109(3):339–345.

Hu C, Zuo H, Li Y, (2021). Effects of radiofrequency electromagnetic radiation on neurotransmitters in the brain. *Front Public Health.* 9:691880. doi: 10.3389/fpubh.2021.691880.

IARC, (2013). IARC monographs on the evaluation of carcinogenic risks to humans. *Nonionizing Radiation, Part 2: Radiofrequency Electromagnetic Fields Lyon*, France, IARC Press. 102: 1–406.

ICNIRP, (2020). Guidelines for limiting exposure to electromagnetic fields (100 kHz to 300 GHz). *Health Phys.* 118(5):483–524.

Khan A, McLeod G, Hidajat T, Edwards EJ, (2023). Excessive smartphone use is associated with depression, anxiety, stress, and sleep quality of Australian adults. *J Med Syst.* 47(1):109. doi: 10.1007/s10916-023-02005-3.

Knott V, Mahoney C, Kennedy S, Evans K, (2001). EEG power, frequency, asymmetry and coherence in male depression. *Psychiatry Res.* 106, 123–140.

Lai H, Levitt BB, (2022). The roles of intensity, exposure duration, and modulation on the biological effects of radiofrequency radiation and exposure guidelines. *Electromagn Biol Med*. 41(2):230–255.

Lefaucheur JP, Aleman A, Baeken C, Benninger DH, Brunelin J, Di Lazzaro V, Filipović SR, Grefkes C, Hasan A, Hummel FC, Jääskeläinen SK, Langguth B, Leocani L, Londero A, Nardone R, Nguyen JP, Nyffeler T, Oliveira-Maia AJ, Oliviero A, Padberg F, Palm U, Paulus W, Poulet E, Quartarone A, Rachid F, Rektorová I, Rossi S, Sahlsten H, Schecklmann M, Szekely D, Ziemann U, (2020). An update (2014–2018). Evidence-based guidelines on the therapeutic use of repetitive transcranial magnetic stimulation (rTMS) *Clin Neurophysiol.* 131(2):474–528.

Lin IC, (1978). *Microwave auditory effects and applications*. Thomas, Springfield, IL.
Lin LC, (2021a). *Auditory effects of microwave radiation*. Springer, Cham.
Lin, JC, (2021b). Sonic health attacks by pulsed microwaves in Havana revisited. *IEEE Microwave Mag*. 22:71–73.
Lin JC, (2022). The microwave auditory effect. *IEEE J Electromagn RF Microw Med Biol*. 6(1):16–28.
Lin-Liu S, Adey WR, (1982) Low-frequency amplitude modulated microwave fields change calcium efflux rates from synaptosomes. *Bioelectromagnetics*. 3:309–322.
Mahamoud YS, Aite M, Martin C, Zhadobov M, Sauleau R, Le Drean Y, Habauzit D, (2016). Additive effects of millimeter waves and 2-deoxyglucose co-exposure on the human keratinocyte transcriptome. *PLoS One*. 11(8):e0160810.
Mallory M, Gogineni E, Jones GC, Greer L, Simone CB, (2016). Therapeutic hyperthermia: the old, the new, and the upcoming. *Crit Rev Oncol Hematol*. 97:56–64. doi: 10.1016/j.critrevonc.2015.08.003.
Meyer F, Bitsch A, Forman HJ, Fragoulis A, Ghezzi P, Henschenmacher B, Kellner R, Kuhne J, Ludwig T, Sachno D, Schmid G, Tsaioun K, Verbeek J, Wright R, (2024). The effects of radiofrequency electromagnetic field exposure on biomarkers of oxidative stress in vivo and in vitro: a systematic review of experimental studies. *Environ Int*. 194:108940. doi: 10.1016/j.envint.2024.108940.
Miller AB, Sears ME, Morgan LL, Davis DL, Hardell L, Oremus M, Soskolne CL, (2019). Risks to health and well-being from radio-frequency radiation emitted by cell phones and other wireless devices. *Front Public Health*. 7:223. doi:10.3389/fpubh.2019.00223.
Mordillo-Mateos L, Dileone M, Soto-León V, Brocalero-Camacho A, Pérez-Borrego YA, Onate-Figuerez A, Aguilar J, Oliviero A, (2017). Effects of transcranial direct current stimulation on temperature and pain perception. *Sci Rep*. 7(1):2946. doi: 10.1038/s41598-017-03173-2.
National Academies of Sciences, Engineering, and Medicine, (2020). *An assessment of illness in U.S. government employees and their families at overseas embassies*. The National Academies Press, Washington, DC. https://doi.org/10.17226/25889.
Neumann WJ, Steiner LA, Milosevic L, (2023). Neurophysiological mechanisms of deep brain stimulation across spatiotemporal resolutions. *Brain*. 146(11):4456–4468.
NIH, (2021). Major depression. https://www.nimh.nih.gov/health/statistics/major-depression
NTP, (2018). National Toxicology Program. Toxicology and carcinogenesis studies in Sprague Dawley (Hsd: Sprague Dawley SD) rats exposed to whole-body radio frequency radiation at a frequency (900 MHz) and modulations (GSM and CDMA) used by cell phones. *Natl Toxicol Program Tech Rep Ser*. Nov;595:NTP-TR-595. doi: 10.22427/NTP-TR-595.
Panagopoulos DJ, Editor, (2022). *Electromagnetic fields of wireless communications: biological and health effects* (1st ed.). CRC Press, Boca Raton, FL.
Perez FP, Maloney B, Chopra N, Morisaki JJ, Lahiri DK, (2021). Repeated electromagnetic field stimulation lowers amyloid-β peptide levels in primary human mixed brain tissue cultures. *Sci Rep*. 11(1):621. doi: 10.1038/s41598-020-77808-2.
Pierpaoli C, Nayak A, Hafiz R, Irfanoglu MO, Chen G, Taylor P, Hallett M, Hoa M, Pham D, Chou Y-Y, Moses AD, van der Merwe AJ, Lippa SM, Brewer CC, Zalewski CK, Zampieri C, Turtzo LC, Shahim P, Chan L, NIH AHI Intramural Research Program Team, (2024). Neuroimaging findings in United States Government personnel and their family members involved in anomalous health incidents. *JAMA*. 331(13):1122–1134.
Rachid F, (2018). Maintenance repetitive transcranial magnetic stimulation (rTMS) for relapse prevention in with depression: a review. *Psychiatry Res*. 262:363–372.

Radzievsky AA, Rojavin MA, Cowan A, Alekseev SI, Ziskin MC, (2000). Hypoalgesic effect of millimeter waves in mice: dependence on the site of exposure. *Life Sci.* 66(21): 2101–2111.

Radzievsky AA, Rojavin MA, Cowan A, Alekseev SI, Radzievsky AA Jr, Ziskin MC, (2001). Peripheral neural system involvement in hypoalgesic effect of electromagnetic millimeter waves. *Life Sci.* 68(10):1143–1151.

Radzievsky AA, Gordiienko OV, Alekseev S, Szabo I, Cowan A, Ziskin MC, (2008). Electromagnetic millimeter wave induced hypoalgesia: frequency dependence and involvement of endogenous opioids. *Bioelectromagnetics.* 29(4):284–295.

Rodina A, Lass J, Riipulk J, Bachmann T, Hinrikus H, (2005). Study of effects of low-level microwave field by method of face masking. *Bioelectromagnetics.* 26(7):571–577.

Romeo S, Sannino A, Rosaria Scarfì M, Lagorio S, Zeni O, (2024). Genotoxicity of radiofrequency electromagnetic fields on mammalian cells in vitro: a systematic review with narrative synthesis. *Environ Int.* 193:109104. doi: 10.1016/j.envint.2024.109104.

Sanders AP, Joines WT, Allis JW, (1985). Effects of continuous-wave, pulsed, and sinusoidal-amplitude-modulated microwaves on brain energy metabolism. *Bioelectromagnetics.* 1985;6(1):89–97.

Sasaki K, Mizuno M, Wake K, Watanabe S, (2017). Monte Carlo simulations of skin exposure to electromagnetic field from 10GHz to 1 THz. *Phys Med Biol.* 62(17):6993–7010.

Schuermann D, Mevissen M, (2021). Manmade electromagnetic fields and oxidative stress-biological effects and consequences for health. *Int J Mol Sci.* 22(7):3772. doi: 10.3390/ijms22073772.

Schuz P, Waldemar G, Olsen JH, Johansen C, (2009). Risks for central nervous system diseases among mobile phone subscribers. *PLoS One.* 4(2):e4389. doi: 10.1371/journal.pone.0004389.

Simko M, Mattsson M-O, (2019). 5G wireless communication and health effects—a pragmatic review based on available studies regarding 6 to 100GHz. *J Environ Res Public Health.* 16:3406. doi: 10.3390/ijerph16183406.

Simkó M, Repacholi MH, Foster KR, Mattsson MO, Croft RJ, Scarfi MR, (2025). Exposure to radiofrequency electromagnetic fields and IARC carcinogen assessment: Risk of Bias preliminary literature assessment for 10 key characteristics of human carcinogens. *Mutat Res Rev Mutat Res.* 796:108545. doi: 10.1016/j.mrrev.2025.108545.

Swanson RL, Hampton S, Green-McKenzie J, Diaz-Arrastia R, Grady MS, Verma R, Biester R, Duda D, Wolf RL, Smith DH, (2018). Neurological manifestations among US Government personnel reporting directional audible and sensory phenomena in Havana, Cuba. *JAMA.* 319(11):1125–1133. doi: 10.1001/jama.2018.1742.

Szabo I, Kappelmayer J, Alekseev SI, Ziskin MC, (2006). Millimeter wave induced reversible externalization of phosphatidylserine molecules in cells exposed in vitro. *Bioelectromagnetics.* 27(3):233–244.

Thair H, Holloway AL, Newport R, Smith AD, (2017). Transcranial Direct Current Stimulation (tDCS): A beginner's guide for design and implementation. *Front Neurosci.* 11:641. doi: 10.3389/fnins.2017.00641.

Thielens A, (2021). Environmental impacts of 5G. A literature review of effects of radiofrequency electromagnetic field exposure of nonhuman vertebrates, invertebrates and plants. Panel for the Future of Science and Technology (STOA). European Parliament, Brussels, European Union; p. 136. EPRS_STU(2021)690021_EN.pdf.

Thomee S, Harenstam A, Hagberg M, (2011). Mobile phone use and stress, sleep disturbances, and symptoms of depression among young adults—a prospective cohort study. *BMC Public Health.* 11:6. doi: 10.1186/1471–2458-11–66.

Vecsei Z, Csatho A, Thuroczy G, Hernadi I, (2013). Effect of a single 30 min UMTS mobile phone-like exposure on the thermal pain threshold of young healthy volunteers. *Bioelectromagnetics*. 34(7): 530–541.

Verma R, Swanson R, Parker D, Ismail A, Shinohara R, Alappatt J, Doshi J, Davatzikos C, Gallaway M, Duda D, Chen HI, Kim JJ, Gur RC, Wolf RL, Grady MS, Hampton S, Diaz-Arrastia R, Smith DH, (2019). Neuroimaging findings in US government personnel with possible exposure to directional phenomena in Havana, Cuba. *JAMA*. 322(4):336–347. https://doi.org/10.1001/jama.2019.9269

Zhadobov M, Chahat N, Sauleau R, Le Quement C, Le Drean Y, (2011). Millimeter-wave interactions with the human body: state of knowledge and recent advances. *Int J Microw Wirel Technol*. 3(2):237–247.

Ziskin MC, (2006). Physiological mechanisms underlying millimeter wave therapy. In: SN Ayrapetyan, MS Markov, editors. *Bioelectromagnetics current concepts. NATO security through science series.* Springer, Dordrecht, pp. 241–251.

8 The limits of health protection and the levels of radiofrequency radiation in a living environment

HIGHLIGHTS

- Electric field strength is an appropriate parameter for describing low-level radiofrequency radiation effects.
- The current guidelines are based on short-term thermal effects and do not consider long-term low-level health effects.
- Nowadays levels of radiofrequency radiation around the world are much lower than recommendations for health protection in the ICNIRP guidelines.
- The total level of radiofrequency radiation decreases with the development of telecommunication technology.
- The level of radiofrequency radiation in the era before mobile telecommunication, 6 V/m, can be considered as a recommended limit of health protection.

8.1 INTRODUCTION

The regulations which are based on guidelines by the International Committee of Non-Ionizing Radiation Protection (ICNIRP), harmonized with the Institute of Electrical and Electronics Engineers (IEEE), are followed in the regulations of radiofrequency radiation by the majority of countries around the world (ICNIRP, 1998, 2020). The ICNIRP guidelines are based on the thermal model and short-term heating. Low-level effects and long-term exposure are not considered by the majority of current regulations.

The thermal model is a threshold model that assumes a linear relation between the effect intensity and the absorbed radiofrequency radiation power, and the specific absorption rate (SAR) is the appropriate indicator of the effect. However, SAR is not singularly determined by the radiofrequency field but depends on the size and electric parameters of the exposed object. The use of SAR in the case of nonthermal effects can lead to confusing results (Section 6.2).

DOI: 10.1201/9781003729976-11

The threshold of the thermal model comes from two studies in groups of eight rats by De Lorge and Ezell (1980) and five monkeys by De Lorge (1984), which showed that trained behavioral skills of the animals decrease, and rectally measured body temperature in monkeys increases by one degree when the SAR exceeds 4 W/kg.

The heating is a cumulative process; the amount of heat increases with the duration of the level. The threshold of the thermal model in the ICNIRP guidelines is described by the SAR at which the temperature in the body rises by one degree in six minutes.

The SAR used in the thermal model is not an independent indicator; it is related to the radiofrequency power density. However, there is no singular relationship between SAR and the parameters of radiofrequency radiation because SAR depends not only on the field parameters but also on the properties of an irradiated object. The correct calculation of SAR, considering the non-homogeneity of the distribution of radiofrequency radiation power density and tissue parameters inside a body, is complicated. Special software and specific anthropomorphic mannequins are required to estimate SAR values. SAR can be useful for the estimation of the level of safety of mobile phones and various other electronic equipment by the producers. SAR can be evaluated in a laboratory, not in environmental conditions. SAR is not a measurable parameter and is not suitable for the evaluation of the impact of radiofrequency radiation in the environment.

The physical mechanism of the non-thermal effect of radiofrequency radiation is related not to SAR but to the electric field strength (Chapter 5). Dielectric polarization, underlying the mechanism, is induced by the electric field and proportional to the electric field strength (Section 1.3). The alterations in the intensity of the effect are proportional not to the alteration of SAR but rather to electric field strength (Suhhova et al., 2013). Describing the possible impact on health, electric field strength is the appropriate parameter for the evaluation of the level of radiofrequency radiation.

The question of whether the impact of microwave radiation is entirely a thermal effect was asked more than half a century ago (Carpenter et al., 1960). The scientists are critical regarding the guidelines, considering only heating and neglecting biological effects detected at the level of radiofrequency radiation much lower than the thermal limits (Barnes & Greenebaum 2020; Lai & Levitt, 2022; Leszczynski & Xu, 2010). During recent decades, the majority of experimental studies have been performed at the levels of radiofrequency levels lower than the ICNIRP guideline's limits. The majority of effects caused by radiofrequency radiation on the human brain, behavior, and cognition were observed at low-level exposure (Section 6.2). Biological and health effects were assessed using the results of studies at low-level exposure (Sections 7.2 and 7.4). Lai and Levitt (2022) concluded that "electromagnetic energy is more biologically potent in causing effects than thermal changes."

Over the last few decades, scientists have proposed various ideas for reducing the threshold of radiofrequency radiation in regulations. A group of scientists, Bioinitiative, proposed to apply a precautionary public health criterion on the minimal level of exposure at detected biological effects. The lowest level for radiofrequency radiation at which an effect was detected was 0.003 μW/cm^2 (0.03 mW/m^2). A reduction from the Bioinitiative 2007 recommendation of 0.001 μW/cm^2 (0.01 mW/m^2) for cumulative outdoor radiofrequency radiation was considered a justified threshold.

Barnes and Greenebaum (2020) proposed an approach that is based on current levels from TV and radio stations and large enough to give signal-to-noise ratios around 20 dB (100-fold) with typical receiving systems. Currently, mean values for the population's exposure to these systems are estimated to be around 0.1 V/m, and peak exposures range up to 2 V/m. The average of 0.1 V/m is based not on research but on the practical need for upgrading current regulations.

Lai and Levitt (2022) proposed another approach that is based on the level at which biological effects occur, considering data from in vivo and in vitro and acute and chronic/repeated-exposure experiments. With an exposure that induces a SAR of 0.0165 W/kg, and using a ten-fold protection, the SAR would be 0.00165 W/kg (i.e., 1.65 mW/kg). For this level, ideas and a thorough justification for the limits are still missing. The main problem is that the proposed SAR limit is applicable in the case of laboratory experiments and electronic devices but not for environmental exposure.

Hinrikus et al. (2023) proposed an idea to use the level of radiofrequency radiation before mobile telecommunication as a basis to which the population is adapted and not causing health problems. Limiting the level of radiofrequency radiation to the peak value of 6 V/m hopefully reduces the health risk to a minimal level and does not restrict the further development of telecommunication technology.

None of these proposals was accepted or implemented in the regulations by the authorities.

This chapter aims to discuss the current situation with radiofrequency radiation in living environments and the criteria for health protection limits, considering both thermal and non-thermal mechanisms. The chapter combines the information about the existing limits and published data about current levels of radiofrequency radiation in different countries to analyze various criteria for updating the limits, involving knowledge from Chapters 5, 6, and 7.

8.2 THE STATE OF LEVEL TO RADIOFREQUENCY RADIATION: THE MEASURED LEVELS OF RADIOFREQUENCY RADIATION

The parameters of radiofrequency radiation in the environment have changed over the years. New frequency bands are being used, and technological developments affect radiation levels. Therefore, the data about the levels of radiation over the recent years are discussed.

The level of radiofrequency radiation alters with the development of services and technology. Total radiofrequency radiation levels in all outdoor areas within only one year increased by about 20–60% in some Belgian cities (Urbinello et al., 2014). The review by Jalilian et al. (2019) does not indicate a noticeable increase in RF-EMF levels since 2012. To understand the current situation, the results of publications on measurements of radiofrequency radiation from recent years are discussed.

Table 8.1 presents an overall picture of outdoor levels of radiofrequency radiation according to measurement data in different countries and regions, showing separately the downlink (which characterizes the radiation level of base stations) and radio broadcasting radiation levels (when possible).

TABLE 8.1

Mean levels of radiofrequency radiation in an outdoor environment

Reference	Countries	Data collected	Total mean	Downlink	Broadcast
Sagar et al. (2018b) Review 24 studies	Austria, Belgium, France, Germany, Greece, Hungary, the Netherlands, Slovenia, Sweden, Switzerland, UK	2000–2015	0.2–0.76 V/m	0.03–0.56 V/m	0.03–0.6 V/m
Jalilian et al. (2019) Review of 26 studies	Belgium, France, Germany, Greece, Hungary, Poland, the Netherlands, Serbia, Spain, Slovenia, Sweden, Switzerland	2015–2018	0.07–1.27 V/m	0.03–1.26 V/m 61–99%	0.03–0.25 V/m
Sagar et al. (2018a)	Switzerland, Ethiopia, Nepal, South Africa, Australia, the United States	2015–2017?	0.23–1.85 V/m	0.47–1.22 V/m	0.02–0.85 V/m
Birks et al. (2018)	Denmark, Slovenia, Spain, the Netherlands, Switzerland	2014–2016	0.076 mW/m^2	0.027 mW/m^2	0.01 mW/m^2
Eeftens et al. (2018)	Denmark, the Netherlands, Slovenia, Spain, Switzerland	2014–2016	0.16; 0.17 mW/m^2	47%, 45 %	25%, 19%
Velghe et al. (2019)	Belgium	2017–2019	9.55 mW/m^2	3.38 mW/m^2	8.88 mW/m^2
	Brussels		2.38 mW/m^2	1.46 mW/m^2	1.29 mW/m^2
	Ghent		9.05 mW/m^2	5.68 mW/m^2	1.86 mW/m^2
	Antwerp		2.9 mW/m^2	1.35 mW/m^2	1.88 mW/m^2
	Hasselt		0.8 mW/m^2	0.73 mW/m^2	0.1 mW/m^2
	Bruges				
Karaesterios et al. (2020)	Switzerland	2002–2016	0.5–2 V/m		
	Slovenia	2005–2011	0.3–5 V/m		
Continuous measurements using telemetric networks	Spain	2005–2020	0.3–5 V/m		
	Greece	2006–2020	0.3–2.8 V/m		
	Germany	2007–2020	0.1–0.6 V/m		
	Ecuador	2012–2016	0.5–5 V/m		
	Hungary	2012–2020	0.5–2 V/m		
	Romania	2014–2020	0.3–5 V/m		
	Serbia	2018–2020	0.3–3.5 V/m		

(Continued)

TABLE 8.1
(Continued)

Reference	Countries	Data collected	Total mean	Downlink	Broadcast
Selmanoui et al. (2021)	South Korea	2019	1.85 V/m Urban zone	3500 MHz 0.14 V/m	
Project 5GEMF1, 2022	Estonia	2022	0.97 V/m	900 MHz 0.64V/m 1800 MHz 0.38 V/m 2100 MHz 0.31 V/m 2600 MHz 0.22 V/m	0.24 V/m
Veludo et al. (2025)	Switzerland	2024	12.08–37.50 mW/m ^{2} Rural area	2100–3500 MHz	

A review by Sagar et al. (2018a) on the distribution of radiofrequency radiation levels in the everyday environment in Europe included the relevant literature published between 1 January 2000 and 30 April 2015. Included in the review were 21 publications that presented data collected between 2005 and 2013. Of the 21 reviewed studies, 10 were spot measurement studies and 11 were personal measurement studies with trained researchers and with volunteers. The measurements were performed in the frequency band from 10 to 6000 MHz. The mean total radiofrequency radiation level for spot measurements in the indoor microenvironment was 0.29 V/m and outdoor 0.54 V/m. In the personal measurement studies with trained researchers, the mean total radiofrequency radiation indoor level was 0.24 V/m and outdoor 0.76 V/m. In the personal measurement studies with volunteers, the mean total radiofrequency radiation indoor level was 0.16 V/m and outdoor 0.20 V/m. In these measurements, the main contributor was the downlink from base stations. The highest mean total radiofrequency radiation value of 1.96 V/m was measured in trains in Belgium, with 95% uplink from the user's device (Sagar et al., 2018b).

The next systematic review on the public level of radiofrequency radiation covered studies published from May 2015 to 1 July 2018 (Jalilian et al., 2019). Twenty-six publications selected for the review included spot and personal measurements in different microenvironments. The measurement devices used in various studies covered the frequencies from 50 to 6000 MHz. The analyses of 26 publications concluded that mean radiofrequency radiation levels in homes, schools, and offices were between 0.04 and 0.76 V/m. Mean outdoor level values varied from 0.07 to 1.27 V/m. The highest levels, up to 1.97 V/m, were measured in public transport stations. Downlink signals from base stations were the dominant contributors (Jalilian et al., 2019).

The international study to assess radiofrequency radiation levels in various microenvironments using portable level meters was performed in six different countries: Switzerland, Ethiopia, Nepal, South Africa, Australia, and the United States (Sagar et al., 2018a). The mean total radiofrequency radiation level in outdoor environments varied between 0.23 V/m (noncentral residential area in Switzerland) and 1.85 V/m (university area in Australia). The mean total RF-EMF level for city centers was 0.48 V/m in Switzerland, 1.21 V/m in Ethiopia, 0.75 V/m in Nepal, 0.85 V/m in South Africa, 1.46 V/m Australia, and 1.24 V/m in the United States, with dominant contribution by downlink level correspondingly 0.47, 0.94, 0.70, 0.81, 0.81, and 1.22 V/m, respectively. The downlink frequency bands 900, 1800, and 2100 MHz were mostly used, except for the LTE band 2600 MHz in the United States. The level of broadcasting radiation was highest at 1.18 V/m in city centers in Australia: Sydney, Canberra, and Wollongong (Sagar et al., 2018a).

The study by Birks et al. (2018) reported the results of three European projects that incorporated five countries: Denmark, the Netherlands, Slovenia, Switzerland, and five regions of Spain. The measurements were conducted between 2014 and 2016 on 529 children. The measurements were conducted at frequencies from 88 to 6000 MHz using the wearable personal level meters for 72 hours and repeated over a year. The total personal level of radiofrequency radiation was 0.076 mW/m^2 (median). Downlink contributed the largest part to the total level (median: 0.027 mW/m^2), followed by broadcast (0.01 mW/m^2). The level from the uplink (0.005 mW/m^2) was lower. The level of WiFi and DECT contributed very little to the total level of radiofrequency radiation (Birks et al., 2018).

Another study on the same projects by Eeftens et al. (2018) was performed on 294 children-parent pairs. The study reported slightly different results: the time-weighted average level was 0.16 mW/m^2 (mean value) for children and 0.15 mW/m^2 for parents. The contribution from downlink sources was 47% for children and 45% for parents, from uplink 18% and 27%, respectively, and broadcast 25% and 19% (Eeften et al., 2018).

The study by Lahham and Ayyad (2019) reported results on a personal level in adults in Palestine. The measurements were performed in the frequency range of 88–6000 MHz during 24 hours in various indoor microenvironments. The average total level was 0.48 V/m, and the maximum varied from 10.2 to 0.9 V/m. The contribution to the mean total level from WiFi 2G was 45%, GSM900 uplink (19%), GSM900 downlink 11%, and FM radio broadcasting 11%. Other sources, GSM1800, UMTS2100, WiFi 5G, DECT, TETRA, WiMAX, and TV bands, all together contributed 14% (Lahham and Ayyad, 2019).

Ramirez-Vazquez et al. (2019) reported the results of the assessment of the personal level of 78 volunteers in Spain. The measurements were performed over 25 hours in various microenvironments (home, walk, work, car, etc.). The total average level of 14 bands in the frequency range from 88 MHz to 5850 MHz was 37.7 μW/m^2, with a minimum of 0.2 μW/m^2 (TV 48.5) and a maximum of 264.7 μW/m^2 (DECT). For Friday, the recorded mean was 53.9 μW/m^2 as opposed to 23.4 μW/m^2 obtained on Saturday. The recorded night-time value was 27.5 μW/m^2 versus 43.8 μW/m^2 recorded in the daytime. The main source that contributed to the mean total personal level was cordless telecommunications (DECT) with 50.2%, followed by mobile phones with 18.4% and mobile stations with 11.0% (GSM, DCS, and UMTS), while WiFi signals gave 12.5%. The total outdoor value was 34.2 μW/m^2.

Velghe et al. (2019) investigated the temporal and spatial variability of radiofrequency radiation in five cities (Brussels, Antwerp, Ghent, Bruges, and Hasselt). The measurements were carried out by two researchers using on-body calibrated personal exposimeters. The frequency was from 50 to 6000 MHz. The highest mean total level 2.63 mW/m^2 was found in Brussels, the most densely populated city in this study. The higher downlink level was measured in Antwerp, which might be an effect of the stronger legislation on base stations in Brussels.

The study by Karastergios et al. (2020) presented the telemetric network aimed at continuous monitoring of radiofrequency radiation and the results of measurements in Greece. Parallel similar telemetric networks in nine other countries were applied: Ecuador, Switzerland, Slovenia, Spain, Greece, Germany, Hungary, Romania, and Serbia. The measurements in the frequency range 100 kHz–7 GHz were conducted every 1 second, and the root mean square values over the 6 minute period were logged to the telemetric network. The level of measured radiofrequency radiation was in the range of 0.1–5 V/m (Karastergios et al., 2020).

The level of radiofrequency radiation for a population living near base stations in France was assessed in the study by De Giudici et al. (2021). The measurements in the frequency range of 100 kHz–6 GHz were implemented in 354 homes in the areas within 250 m of a base station. The sampling intervals of 55 and 150 seconds were chosen for the 48-h and 7-day measurements. The measurements indicated a mean field strength of 0.58 V/m (median 0.44 V/m) for total electromagnetic radiation and 0.43 V/m (0.27 V/m) consequently from the base station radiation. The base stations were a dominant source of radiofrequency radiation in 64% of houses (De Giudici et al., 2021).

The base-station downlink radiation was the main contributor to the total level in many studies. The downlink from the mobile phone base station consists of several bands used by various generations of telecom technology. The results of the studies, as presented in the published papers, do not allow us to differentiate the impact of different generations of telecommunication technology on the downlink radiation.

The level of radiofrequency radiation was measured in 170 locations (larger and smaller cities, settlements in rural areas) in Estonia (605 measurements in total) in January–February 2022 (Project 5GEMF1, 2022). The measurements were carried out in ten different frequency bands: 87–107 MHz, 174–223 MHz, 470–615 MHz, 791–821 MHz, 925–960 MHz, 1805–1880 MHz, 2110–2170 MHz, 2570–2620 MHz, 2620–2690 MHz, 3300–3800 MHz. The total average electric field strength of radiofrequency radiation did not exceed 0.97 V/m (95% confidence interval); the highest level, 1.32 V/m, was measured in urban centers. The average electric field strength in the RadioFM band 87–107 MHz was 0.23 V/m. In telecommunication bands, the level of radiation decreases with frequency: in the band 925–960 MHz (GSM, UMTS) electric field strength was 0.64 V/m, in band 1805–1880 MHz (GSM, LTE) 0.38 V/m, in band 2100 MHz (UMTS) 0.31 V/m, and in band 2600 MHz (LTE) 0.22 V/m (Project 5GEMF1, 2022).

The studies discussed above included frequency bands lower than 2600 MHz, excluding the 5G frequency band, which was not in use yet. The first studies including the 5G at 3500 MHz were performed in South Korea and Switzerland (Selmaoui et al., 2021; Veludo et al., 2025). The measurements of radiofrequency radiation in South Korea were carried out in five large cities and one small city in the area where, in total, 38,999 5G base stations were located. Wideband measurements were carried out in the total frequency band from 80 to 3600 MHz and frequency-selective measurements in the 5G frequency band 3400–3600 MHz. The measured level was highest in dense urban zones, the average level of total radiofrequency radiation was 1850 mV/m, the average level of 5G was 140 mV/m, and the 5G maximum was 2140 mV/m. Averaged over all microenvironment types, the 5G radiation contributed about 15% of total radiation (Selmaoui et al., 2021).

The recent measurements in Swiss including 5G-NR have reported the highest median downlink levels of radiofrequency radiation during the non-user scenario measured in an urban business area of 0.2 V/m (1.02 mW/m^2). Here, downlink and broadcasting bands contributed the most to total radiation levels (Veludo et al., 2025). During the user scenario, the level increased substantially due to the 5G band at 3.5 GHz up to the median level of 2.13 V/m (12.13 mW/m^2), mostly in urban areas. The highest uplink levels were measured in rural environments at 3.76 V/m (37.50 mW/m^2) (Veludo et al., 2025).

The measurement results demonstrate that the radiofrequency radiation levels differ in indoor and outdoor environments. The main contributions to the level of indoor radiation are uplink in transport (metro, trains, buses, and cars, where the radiation propagation conditions are poor due to strong shielding), DECT (cordless telephone), and WiFi at home and WiFi in school-working microenvironments (Lahham and Ayyad, 2019; Ramirez-Vazquezet al., 2019). The radiation from mobile phone base station was the dominant source of level in 64% of houses in urban areas within 250 m of mobile phone base stations (De Giudici et al., 2021).

The maximal total level measured in the radiofrequency range from 10 to 6000 MHz is limited to 2 V/m. The total level of radiofrequency radiation is lower than 2 V/m in all studies except one (Karastergios et al., 2020). The higher level of 5 V/m indicated in the study by Karastergios et al. (2020) is most probably related to the frequency band of 0.1 MHz–6000 MHz, lower than 10 MHz–6000 MHz used in the other studies.

The levels of personal level to radiofrequency radiation presented in the systematic review of studies published in 1998–2021 (56 publications) are much lower than environmental ones (Ramirez-Vazquez et al., 2023). Personal exposimeters were used in 83% of the studies. The lowest mean was measured in Egypt with a value of 0.00100 $\mu W/m^2$ (1.00 nW/m^2) in 2007, and the highest mean was measured in Belgium with a value of 285,000 $\mu W/m^2$ (0.285 W/m^2) in 2019 (Ramirez-Vazquez et al., 2023).

The majority of measurements in Table 8.1 were performed before 5G technology, except for two recent studies considering the impact of 5G at the frequency of 3500 MHz (Selmaoui et al., 2021; Veludo et al., 2025). The measurements performed in South Korea demonstrated that 5G-NR contributed only 15% of total radiation (Selmaoui et al., 2021). The measured total level, including 5G-NR, was limited to 1.85 V/m.

The levels of radiation employed by different generations of telecommunication technology are expected to decrease with the development of technology due to more energy-consuming methods applied (higher sensitivity of receivers, advanced information coding, etc.). The measurements performed in January–February 2022 in

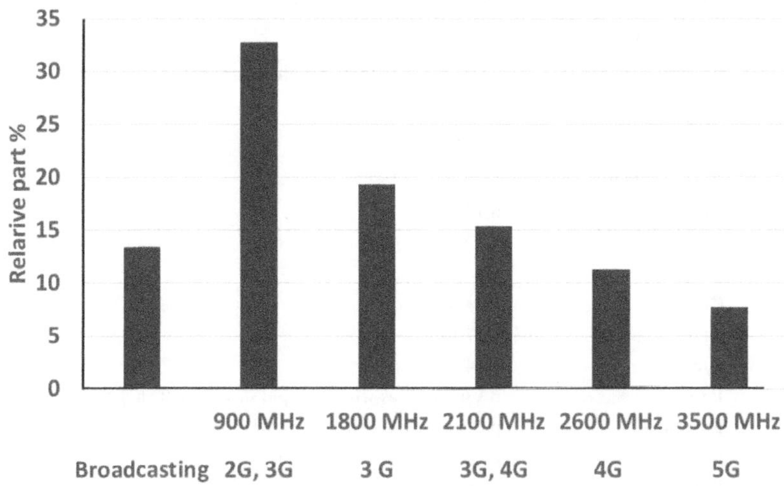

FIGURE 8.1 The relative share of radiofrequency radiation by different telecommunication generations.

Estonia support this assumption (Figure 8.1). Figure 8.1 is based on the measurement data in Estonia in January–February 2022, and the expected level at 3500 MHz is estimated as 15% of the total one according to the conclusion by Selmaoui et al. (2021) (Project 5GEMF1, 2022). The trend of decreasing radiation levels with newer generations of telecommunications is evident. The major part of the radiation is generated by 2G technology at 900 MHz, followed by 3G and 4G.

The results of measurements performed in different countries demonstrate that the actual levels of radiofrequency radiation in the living environment are much below the levels established following the ICNIRP guidelines in the majority of countries.

8.3 CURRENT REGULATIONS OF RADIOFREQUENCY RADIATION IN DIFFERENT COUNTRIES

The majority of countries follow the ICNIRP and the IEEE guidelines and have established a limit of 61 V/m (10 W/m^2) at microwave frequencies (Ramirez-Vazquez et al., 2024). In several countries, the national regulations are much stricter compared to the ICNIRP recommended level (Table 8.2).

The experience of several countries and regions shows that telecommunication systems operate without problems in regions where the electric field strength is much lower than the ICNIRP guidelines. The experience of these countries and regions demonstrates that there are no technological factors determining the need for higher limits and no reasons to apply much higher limits than those recommended by ICNIRP. The changes from previous stricter restrictions to weaker ones following

TABLE 8.2
Countries where the limits of electric field strength are lower than ICNIRP guidelines at microwave frequencies (as of 2021)

Country	Electric field strength
Belgium Brussels	6 V/m
Belgium Flanders	20.6 V/m
Belgium Wallonia	3.0 V/m
Bulgaria	6.1 V/m
Canada	4.7 V/m
Chile	19.4 V/m
China	12.3 V/m
Greece	48.7 V/m
Israel	18.4 V/m
Italy	6 V/m in areas where people stay more than 4 h
Lithuania	previous 19.4 V/m; current 61 V/m
Poland	previous 7.0 V/m; current 61V/m
Russia	previous 6.0 V/m; current 8.7 V/m
Switzerland	6.0 V/m
Turkey	14.4 V/m

ICNIRP recommendations in Lithuania and Poland were political decisions without scientific justification.

The current regulations of radiofrequency radiation levels do not consider the possible long-term exposure to low-level radiation to which the majority of the population is exposed.

8.4 THE IMPACT OF SHORT HIGH-LEVEL PULSES

The possible effect of radiofrequency radiation depends on modulation and on frequency-temporal characteristics of the radiation (Section 5.9). The presence of frequency components close to physiological frequencies in the radiation spectrum is an important factor for the biological effect (Hinrikus et al., 2022).

Telecommunication signals consist of a time series of short pulses. The structure of the telecommunication signals contains the time slots of a fixed number of pulse symbols and frames of a fixed number of time slots determined by the telecommunication standards for the various generations of technology. The transmitter power for a time slot is limited. The standards of 2G, 3G, 4G, and 5G determine the duration of a time slot of 0.5–1 ms at the frequencies lower than 6000 MHz. The modulation frequency components in radiofrequency radiation related to the signal structure of telecommunication systems of various generations are indicated in Table 5.1. The generation beginning from 4G employs a more complex signal structure using orthogonal frequency division multiplexing (OFDM) digital modulation technique, shorter pulses, and time slots than the previous generations 2G and 3G. Therefore, the high pulses cannot appear in older 2G and 3G systems.

The high pulses can be produced due to temporal coincidences between pulses in different channels in the case of multichannel transmission or interference between the beams in various sub-beams in the case of multibeam transmission. In broadband multichannel OFDM systems, interference can take place between subcarriers as well as between symbols (Zhang et al., 2018). The inter-subcarrier interference can be described as the sum of two or more sinusoidal subcarriers. If the subcarriers are simultaneously in phase at some point, the maximum takes place. The amplitude of the maximum is equal to the sum of the amplitudes of the in-phase subcarriers. The level of the maximum increases, and the time interval between them decreases with the increase in the number of in-phase subcarriers. Similar interference between carriers is possible in multiple-input and multiple-output (MIMO) systems.

The time series of short pulse symbols in different channels are independent. By chance, it can happen that the pulses in two or more channels at the same moment simultaneously take their maximal value. As a result of the inter-symbol interference, the pulses can be much higher than the average level of the radiation. Whereas the time series in different channels are random, the concurrency of symbols in many channels is unlikely. Therefore, very high pulses are rare.

The duration of the high peaks is determined by the subcarrier bandwidth and the duration of the OFDM symbols. The maximal duration is 0.067 ms and decreases in broadband 5G systems. The maximal electric field strength (all channels exactly in phase) in a peak is the sum of the electric field amplitudes in different channels. The probability of such a worst-case is very low. More frequently, the fields from only some channels are exactly in phase, and the peak is lower.

Peak-to-average power ratio (PAPR) is used to measure the ratio of peak power to the average power for transmitted radiation. The PAPR for 4G is about 10 dB; for broadband multichannel 5G, the PARP can reach 18–20 dB (Zhang et al., 2018). High PAPR on the input of a receiver presents a serious concern for electronic equipment. Therefore, various methods have been proposed to reduce PAPR and minimize the level of the peaks (Al-Jawhar et al., 2021; Kang et al., 2017). More investigations are needed with broadband 5G to get a better understanding of possible high peaks.

The heating by pulses of durations from 10 to 100 ms at frequencies 6–100 GHz was modeled using the finite-difference time-domain simulation method and bioheat transfer equation (Laakso et al., 2017). The results indicated that the pulsed radiation at the time of average power density limited to ICNIRP guidelines caused a temperature rise of 2° at the frequencies higher than 10 GHz. The increase in temperature rise with frequency can be explained by the fact that the dielectric constant decreases with frequency. Therefore, the reflections from the tissue-air surface decrease, and a larger part of the electromagnetic energy is transferred to the tissue at higher frequencies. The analysis of the averaging effect for level to pulsed radiofrequency radiation has been presented in the study by Neufeld and Kuster (2018). The pulses shorter than 10 seconds at the peak-to-average ratio of 1000 (tolerated by the ICNIRP) can cause tissue damage even after short levels (Neufeld & Kuster, 2018). The authors of these publications concluded that the ICNIRP guidelines may not provide adequate protection from thermal effects at the level of short pulses due to long-level averaging time (Laakso et al., 2017; Neufeld & Kuster, 2018).

The pulses used in broadband multichannel systems are much shorter than the pulses of 10 ms–10 seconds used for the modeling in these studies. The duration of pulses used by the OFDM method varies depending on the number of channels. The MIMO system (4G, 5G, WiFi) uses multipath propagation. The duration of an OFDM symbol is 0.067 ms for 4G and low-frequency 5G (subcarrier bandwidth 15 kHz) and 4 μs for WiFi. For high-frequency 5G, the number of subcarriers and their bandwidth are increasing. The durations of symbols are inversely proportional to the subcarrier bandwidths: 0.067 ms at 15 kHz bandwidth, 0.033 ms at 30 kHz, 0.016 ms at 60 kHz, etc.

The non-thermal model, due to the very small inertia of the process of dielectric polarization, considers the impact of radiofrequency radiation independently of the exposure duration. In principle, even a very short pulse can cause a biological effect. The knowledge about the biological effects of high peaks is still missing. A special design of the experiments is required to evaluate the nature and the impact of high short pulses.

To provide control for health protection conditions, the peak value of the electric field strength should be measured. The limitation of peak values of electric field strength reduces the possible risk related to short pulses. The temporal structure of the generation of high peaks determines the time scale to evaluate the electric field strength: the time should be long enough to detect the appearance of the rear high peaks. To provide reliable detection of the peaks, the monitoring time should be much longer than a time slot. Considering the durations of the time slots, 0.5–1 ms, the time scale of several minutes is relevant for evaluating peak values of radiofrequency electric field strength.

The time averaging in measurements does not allow us to estimate possible short high-level peaks in the radiation. The main contributor to the outdoor radiofrequency radiation is the downlink, where the radiation from 2G and 3G dominates (Figure 8.1). The PAPR is 0 dB for 2G and 3G, where OFDM is not in use. The evaluations of the PAPR for 4G have indicated numbers around 10 dB. At the frequency range 10 MHz–6000 MHz, the PAPR of 10 dB for the total level limited to 2 V/m leads to a peak level lower than 6.32 V/m.

8.5 THE PRINCIPLES AND POSSIBLE CRITERIA FOR HEALTH PROTECTION LIMITS

The linear no-threshold model describes the radiofrequency radiation effects (Section 5.13). The principal threshold of the radiofrequency radiation biological effects is zero. The transfer from biological effects to health impact is a stochastic nonlinear process and has no principal threshold. The zero limit is not applicable in real conditions in everyday life and the environment. A pragmatic approach is needed to solve the problem using alternative possible criteria. The health protection limit can be assessed based on indirect criteria.

The probability of the health effect can be assumed to be very low at the level of radiation lower than the weakest level at which a biological effect has been detected. This approach was used in the proposals by Bioinitiative (2007), Lai and Levitt (2022). Another possible approach, the radiofrequency radiation level is lower than the current radio and TV radiation, was proposed by Barnes and Greenebaum (2020). The third approach, the level of radiation at which the population has lived for a long time without a detectable increase in health risk, was proposed by Hinrikus et al. (2023).

8.5.1 Minimal level of radiation at which the effects were detected

In the majority of experimental studies, only the SAR values have been indicated (including Lai and Levitt, 2022). It is surprising that researchers investigating nonthermal effects use only a parameter of radiation describing the thermal effect. Only a few studies showed the level of the applied radiofrequency radiation field strength and/or power density parallel to the SAR values. The SAR is a parameter determined not only by the level of radiation but also by the parameters of exposed objects. Therefore, the SAR cannot be converted to the field strength or power density of radiofrequency radiation. SAR is not applicable for the evaluation of the level of radiation and cannot be considered as an indicator of the level of the electromagnetic field.

According to the review of neurophysiological effects in the brain in 79 human studies (Section 5.1), the minimal level of SAR at which the effect was detected was 0.003 W/kg, corresponding to the measured parallel electric field strength 2.45 V/m and power density 0.016 W/m^2 (Suhhova et al., 2013).

The review of 112 cellular, human, animal, and plant studies reported a medium SAR value of 0.0165 W/kg. Based on this value, the level of 0.00165 W/kg was proposed as a threshold using a ten-fold protection (Lai and Levitt, 2022). Unfortunately, it is not possible to convert SAR to the radiofrequency field strength or power density of the radiofrequency radiation. The SAR value of 2.1 μW/g (0.00021 W/kg) in the cell culture, indicated as the minimum in Supplement 1 of the review by Lai and Levitt (2022), corresponds to the electric field strength of 17 mV/cm (1.7 V/m) (Velizanov et al., 1999). The study by Eberhardt et al. (2008) indicated a SAR value of 0.00012 W/kg when the rats were exposed to the peak power density of 0.052 W/m^2 (1.4 V/m).

Most likely, the threshold of effect is different in studies performed at different levels, cells, tissues, animals, and humans. Discussing effects in humans, the results of human (maybe animal in vivo) studies can be considered. The minimal levels of the electric field 1.7 V/m causing an effect in a cellular and 1.4 V/m in an in vitro animal study are consistent with the level of the electric field 2.45 V/m causing an effect human study (Hinrikus et al., 2022). In the case of health protection limits, the level of human studies is appropriate. According to these numbers, a level of about 2 V/m can be considered as the possible health protection limit for radiofrequency radiation in humans.

8.5.2 Levels of Broadcasting Radio and TV Radiation in a Living Environment

The level of radio and TV radiation changes with the development of services and technology. In order to implement the ALARA principle, it is necessary to analyze the mutual compatibility of biological and adaptation criteria and the current situation of radio radiation and the development perspective, including the current broadcast radio as a possible starting base. The publications of recent years on radiofrequency radiation measurements characterize the current situation (Section 8.2). The results of measurements of the radiofrequency radiation level in Table 8.1 show that broadcasting radiation contributes significantly to the total level of radiation. The broadcasting radiation constitutes the dominant part, about 93% in some studies (Sagar et al., 2018b; Velghe et al., 2019), 20–45% in the majority of studies (Eeftens et al., 2018; Jalilian et al., 2019; Project 5GEMF1, 2022; Sagar et al., 2018a), and 13% in one study (Birks et al., 2018). The level of total measured radiofrequency radiation is limited to 2 V/m in these studies. Since broadcasting contributes a considerable part of that, the level of broadcasting radiation is comparable to that, about 1–2 V/m. The level of 1 V/m can be considered as a possible health protection limit. This level is somewhat higher than that considered by Barnes and Greenebaum 2020 in their proposal for radiofrequency limits of 0.1 V/m.

8.5.3 The Level of Radiofrequency Radiation before the Mobile Telecommunication Era

Mankind has been living in radiofrequency radiation for about a century, and no special health complaints related to it have been noticed before mobile telecommunication and microwave radiation. The level of radio broadcast radiation before mobile telecommunication can be considered as a reference level for the health protection limits.

The character of historical broadcast radiofrequency radiation differed from modern telecommunication radiation: the used frequencies were lower, and amplitude-modulated (AM) or phase-modulated (FM) signals were used instead of digital signals in modern telecommunication. The sound frequencies are close to physiological frequencies; therefore, the impact of broadcast radio was stronger. The impact of radiofrequency radiation decreases with frequency due to decreasing dielectric polarization and the depth of penetration (Chapter 3). The fact that no increase in health risk has been revealed for the population living in broadcast radiofrequency fields over decades can be explained by the ability of natural physiological mechanisms to repair the consequences possible at that level of radiation. People are adapted to living at low-level to broadcast radiofrequency radiation.

In 1979, the measurements were performed in 15 large cities in the United States in frequency bands 54–88 MHz (TV), 88–108 MHz (FM Radio), 174–216 MHz (TV), and 410–806 MHz (TV) (Tell and Mantiply, 1980). The results of the measurements indicated a median level 5 mW/m^2. The authors reported that 99% of the population was potentially exposed to radiation at levels lower than 100 mW/m^2 (6 V/m).

The level 6 V/m was measured in radiofrequency bands lower than microwaves. At higher frequencies the impact of radiation decreases due to decreasing dielectric permittivity (Section 3.2). Therefore, the limit determined at lower radiofrequency is even stronger at higher microwave frequencies.

The limit of 6 V/m is the most conservative one and weaker compared to the limits determined by the minimal level of the radiofrequency radiation at which the effect has been detected in human experimental studies, 2 V/m, or the current level of broadcasting, about 1 V/m. The limit of 6 V/m is close to the regulations in some countries and regions (Brussels, Bulgaria, Canada, Italy, Switzerland). The experience of these countries demonstrates that this level is not an obstacle to current telecommunication services. The total limit of 6 V/m is even higher compared to the current environmental level of microwaves, limited to 2 V/m (Table 8.1). Since the development of technology has decreased the level of radiation in telecommunication systems, the level of 6 V/m cannot prevent the development of technology. This limit is so conservative that it does not change the current general situation in microwave frequencies (Table 8.1). However, the limit of 6 V/m will regulate the output power of transmitters and distances between base stations, limiting the level of radiation in the areas close to antennas.

All discussed criteria suggest the level of health protection limits much lower than the ICNIRP criterion 61 V/m (10 W/m^2).

8.6 CONCLUSIONS

The analysis of possible approaches for limiting exposure to microwave radiation, considering the current situation with environmental exposure to radiofrequency radiation, leads to the following conclusions.

- The current environmental level of radiation in microwave frequencies does not exceed 2 V/m worldwide.
- The contribution from broadcasting radio radiation dominates the total level of radiofrequency radiation.
- The level of radiofrequency radiation, 6 V/m, in the era before mobile telecommunication can be considered as an appropriate limit of health protection in microwave frequencies.
- The limit of 6 V/m does not affect the current general situation and does not hinder the development of telecommunication technology.
- Despite the proposed level being higher than the level of radiation causing an effect in humans at 2.45 V/m, the actual level does not exceed 2 V/m.
- Following the precautionary principle is important for minimizing the risk of health effects.

REFERENCES

Al-Jawhar YA, Khairun NR, Taher MA, Shah NSM, Mostafa SA, Khalaf BA, (2021). Improving PAPR performance of filtered OFDM for 5G communications using PTS. *ETRI Journal.* 43(2):209–220.

Barnes F, Greenebaum B, (2020). Setting guidelines for electromagnetic levels and research needs. *Bioelectromagnetics.* 41(5):392–397.

Bioinitiative, (2007). Bioinitiative report: a rationale for biologically-based exposure standards for low-intensity electromagnetic fields. Bioinitiative 2012. Assessed 17/06/ 2025. https://bioinitiative.org/search/2007+report/

Birks LE, Struchen B, Eeftens M, van Wel L, Huss A, Gajšek P, Kheifets L, Gallastegi M, Dalmau-Bueno A, Estarlich M, Fernandez MF, Meder IK, Ferrero A, Jim´enez-Zabala A, Torrent M, Vrijkotte TGM, Cardis E, Olsen J, Valič B, Vermeulen R, Vrijheid M, Röösli M, Guxens M, (2018). Spatial and temporal variability of personal environmental level to radio frequency electromagnetic fields in children in Europe. *Environ Int.* 117:204–214.

Carpenter RC, Biddle DK, Van Ummersen TA, (1960). Opacities in the lens of the eye experimentally induced by level to microwave radiation. *IRE Trans Med Electronics.* ME7:152–157.

De Giudici P, Genier JC, Martin S, Doré JF, Ducimetière P, Evrard AS, Letertre T, Ségala C, (2021). Radiofrequency level of people living near mobile-phone base stations in France. *Environ Res.* 194:110500. doi: 10.1016/j.envres.2020.110500

De Lorge JO, Ezell CS, (1980). Observing-responses of rats exposed to 1.28- and 5.62-GHz microwaves. *Bioelectromagnetics.* 1(2):183–198.

De Lorge JO, (1984). Operant behavior and colonic temperature of Macaca mulatta exposed to radio frequency fields at and above resonant frequencies. *Bioelectromagnetics.* 5(2):233–246.

Eberhardt JL, Persson BR, Brun AE, Salford LG, Malmgren LO, (2008). Blood-brain barrier permeability and nerve cell damage in rat brain 14 and 28 days after level to microwaves from GSM mobile phones. *Electromagn Biol Med.* 27(3):215–229.

Eeftens M, Struchen B, Birks LE, Cardis E, Estarlich M, Fernandez MF, Gajšek P, Gallastegi M, Huss A, Kheifets L, Meder IK, Olsen J, Torrent M, Trček T, Valič B, Vermeulen R Vrijheid M, van Wel L, Guxens M, Röösli M, (2018). Personal level to radio-frequency electromagnetic fields in Europe: is there a generation gap? *Environ Int.* 121:216–226.

Hinrikus H, Lass J, Bachmann M, (2021). Threshold of radiofrequency electromagnetic field effect on human brain. *Int J Radiat Biol.* 97(11):1505–1515.

Hinrikus H, Koppel T, Lass J, Orru H, Roosipuu P, Bachmann M, (2022). Possible health effects on the human brain by various generations of mobile telecommunication: a review based estimation of 5G impact. *Int J Radiat Biol.* 98(7):1210–1221.

Hinrikus H, Koppel T, Lass J, Roosipuu P, Bachmann M, (2023). Limiting level to radiofrequency radiation: the principles and possible criteria for health protection. *Int J Radiat Biol.* 99(8):1167–1177.

ICNIRP, (1998). Guidelines for limiting level to time-varying electric, magnetic and electromagnetic fields (up to 300 GHz). *Health Phys.* 74(4):494–522.

ICNIRP, (2020). Guidelines for limiting level to electromagnetic fields (100 kHz to 300 GHz). *Health Phys.* 118(5): 483–524.

Jalilian H, Eeftens M, Ziaei M, Röösli M, (2019). Public level to radiofrequency electromagnetic fields in everyday microenvironments: An updated systematic review for Europe. *Environ Res.* 176:108517.

Kang C, Liu Y, Hu M, Zhang H, (2017). A low complexity PAPR reduction method based on FWFT and PEC for OFDM systems. *IEEE Trans Broadcast.* 63(2):416–425.

Karastergios I, Gialofas A, Karabetsos E, (2020). National observatory of electromagnetic fields: national telemetric network for the measurement of high-frequency electromagnetic fields in Greece. *Radiat Prot Dosimetry.* 188(4):413–423.

Laakso I, Morimoto R, Heinonen J, Jokela K, Hirata A, (2017). Human level to pulsed fields in the frequency range from 6 to 100 GHz. *Phys Med Biol.* 62(17):6980–6992.

Lahham A, Ayyad H, (2019). Personal level to radiofrequency electromagnetic fields among Palestinian adults. *Health Phys.* 117(4):396–402.

Lai H, Levitt BB, (2022). The roles of intensity, level duration, and modulation on the biological effects of radiofrequency radiation and level guidelines. *Electromagn Biol Med.* 41(2):230–255.

Leszczynski D, Xu Z, (2010). Mobile phone radiation health risk controversy: the reliability and sufficiency of science behind the safety standards. *Health Res Policy Syst.* 8(1):2. https://health-policy-systems.biomedcentral.com/articles/10.1186/1478-4505-8-2

Neufeld E, Kuster N, (2018). Systematic derivation of safety limits for time-varying 5G radiofrequency level based on analytical models and thermal dose. *Health Phys.* 115(6):705–711.

Project 5GEMF1, (2022). Assessment of current and 5G caused possible health effects related to nonionizing radiation. Final Report, Tallinn University of Technology, Estonia. https://www.sm.ee/uudised/uuring-uuemate-4g-ja-5g-polvkondade-tehnoloogia-kasutuselevotuga-kiirguse-tase-langeb

Ramirez-Vazqueza R, Gonzalez-Rubiob J, Arribasa E, Najerab, (2019). Characterisation of personal level to environmental radiofrequency electromagnetic fields in Albacete (Spain) and assessment of risk perception. *Environ Res.* 172:109–116.

Ramirez-Vazquez, R., Escobar, I., Vandenbosch, G.A.E., Vargas, F., Caceres-Monllor, D. A., Arribas, E, (2023). Measurement studies of personal level to radiofrequency electromagnetic fields: a systematic review. *Environ Res.* 218:114979. doi: 10.1016/j.envres.2022.114979.

Ramirez-Vazquez R, Escobar I, Vandenbosch GAE, Arribas E, (2024). Personal level to radiofrequency electromagnetic fields: a comparative analysis of international, national, and regional guidelines. *Environ Res.* Apr 1;246:118124. doi: 10.1016/j.envres.2024.118124.

Sagar S, Adem SM, Struchen B, Loughran SP, Brunjes ME, Arangua L, Dalvie M A, Croft RJ, Jerrett M, Moskowitz JM, Kuo T, Roosli M, (2018a). Comparison of radiofrequency electromagnetic field level levels in different everyday microenvironments in an international context. *Environ Int.* 114: 297–306.

Sagar S, Dongus S, Schoeni A, Roser K, Eeftens M, Struchen B, Foerster M, Meier N, Adem S, Roosli M, (2018b). Radiofrequency electromagnetic field level in everyday microenvironments in Europe: a systematic literature review. *J Eposure Sci Environ Epidemiol.* 28 (2):147–160.

Selmaoui B, Mazet P, Petit P-B, Kim K, Choi D, de Seze R, (2021). Level of South Korean population to 5G mobile phone networks (3.4–3.8 GHz). *Bioelectromagnetics.* 42(5):407–414.

Suhhova A, Bachmann M, Karai D, Lass J, Hinrikus H, (2013). Effect of microwave radiation on human EEG at two different levels of level. *Bioelectromagnetics.* 34:264–274.

Tell RA, Mantiply ED, (1980). Population level to VHF and UHF broadcast radiation in the United States. *Proceedings of the IEEE.* 68(1):6–12.

Urbinello D, Joseph W, Verloock L, Martens L, Roosli M, (2014). Temporal trends of radio-frequency electromagnetic field (RF-EMF) level in everyday environments across European cities. *Environ Res.* 134:134–142.

Velghe M, Joseph W, Debouvere S, Aminzadeh R, Martens L Thielens A, (2019). Characterisation of spatial and temporal variability of RF-EMF level levels in urban environments in Flanders, Belgium. *Environ Res.* 175:351–366.

Velizarov S, Raskmark P, Kwee S, (1999). The effects of radiofrequency fields on cell proliferation are non-thermal. *Bioelectrochem Bioenerg.* 48(1):177–180.

Veludo AF, Stroobandt B, Van Bladel H, Sandoval-Diez N, Guxens M, Joseph W, Röösli M, (2025). Exploring RF-EMF levels in Swiss microenvironments: an evaluation of environmental and auto-induced downlink and uplink level in the era of 5G. *Environ Res.* 266:120550. doi: 10.1016/j.envres.2024.120550.

Zhang L, Ijaz A, Xiao P, Molu M, Tafazolli R, (2018). Filtered OFDM systems, algorithms and performance analysis for 5G and beyond. *IEEE Trans Commun.* 66(3):1205–1218.

Index

Note: **Bold** page numbers refer to tables; *italic* page numbers refer to figures.